Smitten by Gold

The Highs and Lows of Amateur Gold Panning

It is said, 'gold is where you find it'. On this particular occasion, it came straight out of a vial. This is one of a series of photos taken of the author for a booklet on how to use a pan, hence the Indiana Jones hat. It is never used while on a serious day's panning.

Text © Alan C. Souter, 2020.
First published in the United Kingdom, 2020,
by Stenlake Publishing Ltd.,
54-58 Mill Square,
Catrine, Ayrshire,
KA5 6RD
Telephone: 01290 551122

www.stenlake.co.uk

Printed by Claro Print Ltd.,
Offices 26/27,
1 Spiersbridge Way,
Thornliebank, Glasgow,
G46 8NG

ISBN 9781840338799

**The publishers regret that they cannot supply
copies of any pictures featured in this book.**

Acknowledgements

I am indebted to a lot of people for their help in getting this book to fruition. Leon Kirk started the process but continued to support on a regular basis. Paul Jacobs, the professional photojournalist, has allowed me to use his photos gratis. Leon Kirk and Charlie Smart have also donated photos. Julian Bagg kindly agreed to my use of his cartoon.

To Richard Stenlake and Callum Johnston of Stenlake Publishing Limited I owe a big debt of gratitude for not only agreeing to publish my raw text, but for editing and crafting the final form.

A large number of people spent time telling me their stories. Some were good enough to write me at length or sit for an interview. I must, however, single out a number by name for their extensive contribution. Vince Thurkettle, John Greenwood, John Hooper, Alf Henderson, Bob Sutherland, Kit Andrews, Andrew Winter, Mark Oddy, Dave Jones, Mike Jones, Robbie Falconer and Stan Johnston. To those and others from whom I gleaned material, I give my thanks.

I appreciate Dr. Neil Clark's help in checking technical data.

I would like to thank a number of landowners for allowing me to practise my hobby in their streams, but I will not name them for fear of starting a mini rush. However, the late Mr. EM Reeves and his daughter Lucy deserve special mention for their enlightened policy of allowing me and hundreds of others to pan in the rich gravels of Suisgill estate streams when in their ownership.

Finally many of the anecdotes documented in these pages have come when out with friends and acquaintances. I cannot recall an unenjoyable day. I thank you for your friendship and shared experiences. I will not make a list of all those who have made looking for gold so pleasurable, you know who you are.

Dedication

This book is dedicated to my wife, Colinne, who rather than be a 'gold panner's widow' quickly learned to pan and accompanied me on numerous enjoyable prospecting trips at home and abroad.

Preface

This book has been a long time in the writing. The idea began with a conversation between Vince Thurkettle and Leon Kirk. Thoughts were voiced regarding the stories concerning gold panners in Britain being lost as they, and those who knew them, passed on. There was a need for someone to record them. Leon had seen my diary and approached me to write something. Having retired, I did have a little more time, but over the years I had done my prospecting alone for the most part. This meant I was not privy to the conversations of the main body of panners and although I had bumped into and spoken briefly to many, I was not in a position to fulfil the main aim.

To rectify that, I joined the British Gold Panning Association in 2011 and went along to the annual panning championships in Wanlockhead. From then on I began filling notebooks with stories and incidents. I made the odd recording, but frequently I found that the best stories were related when I was least prepared. This resulted in a later rough outline of the story. Some of these I was able to follow up and get detail, but others remained in a less precise state. I apologise if a reader with links to the story feels that my account is not exactly as it happened.

The project did not develop as I had initially intended. I realised I was not getting all the stories I would like, but those I did get bore a great resemblance to my own experiences. I therefore changed tack to give it more of a biographical framework onto which I could latch other stories. With each passing year a crop of new stories was being added and it became apparent that I could go on collecting and recording for years. This was particularly so as the idea of large group meetings (digs) took hold and post-digging chat elicited a wealth of tales. As the World Gold Panning Championships in Moffat in 2017 approached, I felt it was time to wind up the process and finish adding to the text.

I am grateful to those who responded directly and provided their stories in their own words. Rather than rewrite them, I have incorporated most of them verbatim as I feel it adds to the colour (excuse the pun).

Alan C. Souter
Aberlour, 2020

IN THE BEGINNING

'I know what gold does to men's souls.'
Howard – the old prospector in *Treasure of the Sierra Madre*

Why do people get gold fever? There is little doubt that the early gold rushes were driven by people who hoped to 'get rich quick'. By the time I was born almost all the large gold fields had been discovered, yet like many, I contracted gold fever and it has no known cure. My wife has accused me of speaking about nothing else, but it is the pastime from which I derive greatest pleasure and as other interests have been slowly abandoned, my time spent looking for gold has increased. Many can't give up golf or fishing. I've given up both, but not the desire to find gold. I have a French friend who is passionate about salmon angling. 'Alan, fishing is my drug' he says. Looking for gold is every bit as addictive and probably more so.

There are still those in this country who hear about gold finds and think they will make a bit of money. There are few who, if they tried balancing the books, would find that their income exceeded their expenditure on equipment alone. I used to respond to enquiries from those looking to find out the worth of what I showed them by saying 'It won't cover the beers'. In truth, I have several thousands of pounds worth, but every speck, flake and nugget is still in my possession in some form or other.

If looking for gold is not about its monetary value, what is it about? My feelings about this obsession is summed up in the lines of the last verse of Robert Service's poem *The Spell of the Yukon*:

> There's gold, and it's haunting and haunting,
> It's luring me on as of old;
> Yet it isn't the gold that I'm wanting
> So much as just finding the gold.
> It's the great, big, broad land 'way up yonder,
> It's the forests where silence has lease;
> It's the beauty that thrills me with wonder,
> It's the stillness that fills me with peace.

The phrase 'just finding the gold' involves a lot. Like most gold seekers, I get a buzz from seeing that first glint in the pan or on rock, particularly, but not solely from a burn I have not tried before.

Vince Thurkettle has been quoted in print as follows. "I get a thrill every time I find any speck of gold because to me it's a treasure hunt with nature". I know a lot of British panners feel the same way.

Programmes like The Discovery Channel's *Gold Rush Alaska* and *Gold Divers* have raised the awareness that gold can still be found by newcomers. From time to time, the press in Britain rework the old cliché 'There is gold in them thar hills'. So the 21st century's boom in amateur panners and prospectors has been fuelled by the media.

But what of those who have been looking for gold in Britain for decades? What made them start? It can't be in the genes. Or is it? I discovered that my great great grandfather, John Hunter Bain, left Scotland, his wife and family and went to the Australian gold rush, later dying in Dunedin, New Zealand. His profession was given as moulder, so more than likely it was the hope of financial improvement and perhaps a sense of adventure that made him go.

The author as a student digging holes for geomorphological research.

For my part, as I've already explained, the financial side has never been the motive. Like some other panners, I enjoy finding shiny red lumps of hematite and will often take home several samples from streams like the Mennock and Wanlock waters. But finding hematite samples does not develop into the passion which is fuelled by finding very much smaller specks of gold. So there has to be some other explanation. Gold panning is done because it gives a high. It also has its lows. These are usually forgotten almost immediately by those with gold fever. A new wave of optimism fires the desire to go on another trip. The fever can be alleviated temporarily by finding gold, but can never be fully cured. Reading about or watching a TV programme on gold is guaranteed to make the symptoms worse. Nonetheless, if there is one thing I've learned about the whole gold seeking business, it is that no matter how many times I go looking, I never stop learning.

This book is about my experiences and those of various gold seekers past and present. If you have already found your first gold, you will not need this book to be itching to get back to a stream in the near future but, hopefully you will also be armed with a few more ideas, at the same time avoiding the pitfalls which almost every gold panner has suffered. If you have yet to find your first gold, there will be plenty of information to get you started, but be warned, to parody Leon Kirk's words on his Gold Panning Supplies website…

> …I 'cannot be held responsible for you getting gold fever'.

In his day, Bob Sutherland from Montrose was one of Britain's best gold panners. At the 2011 British Gold Panning Championships in Wanlockhead, I asked Bob, by then a spectator in all things gold panning, how he got started. Living in Montrose near the agate filled lavas of the Lower Old Red Sandstone he had developed an interest in minerals. By a twist of fate this had led to looking for gold. Once he was smitten with gold fever, the other minerals took a back seat. My path to gold digging took a similar, if more prolonged course.

My primary schooling was in the small Banffshire town of Portsoy. There, I had the freedom to roam the coast from Sandend in the west to Boyne Bay in the east. At the same time I looked out on the Durn Hill with its orange quartzite quarry.

One of the earliest games I played with friends was to mine a soft rock on the west braes using old bits of bicycle washed up on the beach near the town's rubbish dump. I enjoyed digging. At one scout camp I was given the job of digging the latrines. We happened to be in an area of sandy deposits (probably fluvio glacial) and the excavation went well. I had determined to make a pit that would last the week. When completed, I was helped out and a sitting bar was duly erected with a canvas screen around.

Early next day, calls for help were heard from the structure. One of the troop with a considerably smaller stature, had overbalanced and was trapped like a bear in a pit. My hole digging career had begun.

Portsoy gave me my first passion, fishing. The Durn Burn was where I caught my first brown trout and later I became accomplished at catching the short, fat sea trout of the Boyne Burn. One of the best pools was below Boyne Castle and I spent many an hour on the rock promontory at the head of the Castle Pot, trotting a worm down the current. It is one of these places that recalls a particular tune. Mantovani's melody of the moment was *Charmaine*. If the sea trout had not yet run, I would visit the mouth, climb the brae beside the limestone quarry and watch the fish jumping.

Everyone in Portsoy knew that 'Portsoy Marble', the serpentine found west of the harbour, had been used to make a fireplace in the Palace of Versailles. If you found a pebble on the beach with a vein through it, it could be cut with a hacksaw, polished with wet and dry paper, drilled and turned into a pendant.

At Redhythe to the east the red rocks had large lumps of shiny mica and long black crystals which I was told was tourmaline. Later, on going through the pages of Heddle's *A Mineralogy of Scotland* I discovered numerous references to the minerals of Portsoy and I continued to visit to build a collection, even after moving west to live in Buckie.

My geography teacher at Buckie High School inspired me. Bob Denholm not only helped me catch my first sea trout on the fly, he also gave me a love of physical geography. This led to applying to Aberdeen University, where I eagerly signed up for both the geology and geography courses and another two sciences, one of which was Zoology. I thought Zoology would be appropriate because I went bird watching with my father. The zoology professor, Prof. Wynne Edwards, had studied the rookery in Old Aberdeen, so I thought there would be plenty of interest. In the event, the one lecture on the rook colony could not make up for the tedium of numerous hours dissecting worms, rats and dog fish. If there was one benefit, it was that I had had to purchase a dissection kit. In later life the fine pointed scissors became very useful for fly tying but the tweezers now travel in my box of gold vials into which each day's take is decanted. They are also the backup, though rarely used tool for pulling nuggets from cracks.

On the other hand, geology field excursions were a great pleasure and in December 1963 I borrowed money from my girlfriend to go on the Geology Society's trip to the Isle of Rhum. I took a pan with me. This was an aluminium frying pan from which I had removed the metal handle. There, I hoped to start my gold panning career. Had I done a little more research beforehand, I would have realised the gabbro and ultrabasic rocks around Allival and Askival were very unlikely territory. The lecturer leading the group, Bill Welch, left me in no doubt that I had a lot to learn about mineralogy.

At university, my hole-digging talents were put to further use by a young research student, John Smith later to become Professor John Smith. I had volunteered to go to Tarradale House near Muir of Ord, to help with its refurbishment as a field centre. When my girlfriend arrived, she was told I was out on the terrace. Thinking that I would be on a patio area at the front of the house, she was surprised that I was nowhere to be seen. She then recalled the term terrace meant something totally different to a geographer and geologist. I was barely visible on the level overlooking the mouth of the River Beauly, having cut a section through the sand and gravel layers to a depth of around five feet. There was no thought of gold in this dig, though in later years, I did dig profitably to a similar depth below water level in the gravels of the Kildonan.

Digging deep into the Kildonan gravels. The drysuit was an eBay buy for digging this specific spot.

Years later, I had displayed in my classroom, a collection of rocks and minerals from all over the north of Scotland and further afield. A young pupil approached me expressing an interest. He said he had a relation working in Kildonan and that there was gold there. I had founded the 'Outdoor Club' at Banff Academy and at every opportunity I would take pupils to the hills to either walk or ski. In the Cairngorms we had picked up the odd bit of smoky quartz or cairngorm crystal, but here was a chance to run a slightly different trip. Contact was made with the factor of the Suisgill Estate and a trip was arranged. On arrival he said the easiest way to find gold was to "moss wash".

Despite gold's high specific gravity and its tendency to drop down through water faster than almost all other minerals, fine gold can get lifted in a flood and carried over moss covered rocks. There, it gets filtered out and concentrated in the moss. This is the process that has been replicated by using carpet or miner's moss in gold sluices. I know of one spot on the Kildonan where the river narrows between rock walls and where the level in a big flood rises a lot higher than in areas where it has room to spread sideways. There is a small bay in the rock covered in moss and wood-rush. Here the velocity of the water slows and the heavier minerals get trapped by the vegetation. The 'soil' has an extremely high black iron content and in it there are numerous tiny pieces of gold. This spot can give freshly deposited flood gold seven or eight feet above normal water level. In the sixties, the moss nearer normal water level was full of pieces up to a couple of millimetres across. It was duly stripped off the rock and thoroughly washed in our frying pans. There was no doubting that the little shiny pieces appearing out of the black iron sand were specks of gold. They went into a small clear aspirin bottle and by the time we had to leave, it contained a visible line. I have kept this gold in a separate vial to this day.

This trapping of gold by fibres is the process which gave rise to the Greek legend of Jason's Golden Fleece. Jason and the Argonauts headed for Colchis in the eastern Black Sea. It is now part of the republic of Georgia. Examples of sophisticated bronze working from this area predate its development in Europe and exquisite gold jewellery artefacts date back to around the 5th century BC. Some of the rivers and streams flowing from the Caucasus to the north contain gold. In the last century the villagers in the Svaneti area were observed using sheep fleeces pegged to the river bottom to filter out the alluvial gold from the moving gravels transported by the rivers swollen by snow melt. Once the fleece was dried the gold could be shaken out. Was this what gave rise to the Golden Fleece legend? Who knows, but I remember decades ago seeing carpet held down by large stones on the bed of the Kildonan. Had the idea spread?

> *Fact*: There are around 1.3 grams of gold per 1,000 tonnes of other material in the crust.

While I was moss washing, others with more experience, were digging the gravels. On one walk up the Suisgill I met a small party who showed me a much larger medicine bottle almost full of very chunky pieces. By the time I had time to spend more than a day or two a year, the Suisgill had been closed to panners.

Although my introduction to the pastime was drawn out and relatively unproductive, I know of one case where a very well-known panner had an epiphany.

Vince Thurkettle is probably Britain's best-known panner. He was president of the World Goldpanning Association for seven years and part of the British gold medal winning team at the World Goldpanning Championships in Canada in 2007. I leave Vince to tell his own story:

My introduction to gold was sudden and unexpected. In the late 1970s I was at college studying forestry and had developed a particular liking for the local Lakeland geology. This rapidly evolved through rocks to minerals and from scavenging over old mine spill heaps to abseiling down the shafts of abandoned mines looking for fine crystals and ore specimens. While underground, night and day make no difference, so on this particular life-changing night I set off just after midnight with two friends to explore an old mine, hoping to find some crystals of scheelite; an attractive ore of tungsten.

Sometime around 1.30, about a kilometre underground and with just enough scheelite to be pleasing, we prepared to return home. A slight echo startled us and we looked back into the access tunnel and were horrified to see two miner's lights approaching. I fleetingly considered fleeing deeper into the mine, but decided that would be really stupid and that I would just have to take my punishment. We had been caught red-handed deep in a mine without permission – I stood up and walked towards the approaching lights, the sooner I got the inevitable scolding over the better.

But as I reached them the stranger's words were hesitant and friendly and not at all the authoritarian barrage I had been expecting. The whole tense situation suddenly melted into a huge joke as it turned out that they had no permission to be in the mine either and had thought that we might be the miners. We all left the mine together just before dawn and the strangers invited us back for coffee – a chance to meet properly and discuss such things as apatite, scheelite and wolframite. It is a sadness to me that I neither remember the face nor name of the man with whom I had coffee that morning. He was interesting and a passionate mineralogist. He had a fine cabinet well stocked with mineral specimens that he had collected, but what caught my eye as being the most unusual, was a row of clear plastic blocks each containing a small sample of natural gold. In my developing passion for hunting minerals it had never occurred to me that it might still be possible to go into the wilds and find gold. Guardedly, as in my heart I distrusted what my eyes were telling me, I asked him about the gold specimens. I had the prejudice, still quite common in many people today, that, while gold was once plentiful and could readily be found during the Victorian period, that time was over now and the gold is all gone. I sipped my coffee and listened with a narrowing fascination, I was by now oblivious to every other beautiful crystal and brightly coloured mineral in the cabinet – he talked, I stared at the gold, and the gold smiled back at me! In the space of barely 15 minutes I had had the quasi-religious experience of complete conversion – the complications and attractions of schists, carbonates, calcites and chlorides fell away, for me now there was only one goal, to find my own specks of the singular, star-born element – gold!

The summer holidays were approaching, but I couldn't wait to get started and the next Sunday morning found me sneaking behind the college kitchens to examine the dustbins. The bins were of heavy metal, with solid goldpan-shaped metal lids, they would do fine I thought, and stole two. A friend and I spent all that day hopelessly sloshing gravel around in our dustbin lids in the clear waters of a stream near the Saddleback Mountain – we found absolutely nothing, but it was a first step and I was undaunted.

The term ended and I trudged home to my parents, mildly irritated at having come last in my final forestry exams. I had been puzzled by this and had broken into the Head of the Forestry Department's office to review the marking of my exam papers. Annoyingly they had been fair. I had written an extremely poor paper on the use of explosives in forest road building – the problem was that the question had actually asked about machinery use, I had just thought that it would make my answer more interesting if I livened it up a bit with explosives – it hadn't. Shaking off the spectre of the pitiful result of a three-year college education, I set about finding a gold pan.

In the days before the Internet, life was in many ways more interesting. Knowledge was a precious thing residing in people and libraries – it took careful detective work and effort to learn something new, such as, where to buy a gold pan in England? I couldn't afford to ship a metal pan over

Vince Thurkettle and Mark Gregory with a dilemma.

from America and my enquiries had met a complete blank and a good deal of sarcastic mirth. I was in a narrow street in Norwich of all places, when I noticed a shop with an unusually wide range of peculiar gift items. I entered and with a deadpan expression asked a tall man in his 40s if he sold gold pans, I then paused, waiting for some variation of the usual ridicule that followed the question, but he smiled. "Of course!" he said. I left the shop beaming, with two simple 14 inch metal gold pans. The shopkeeper had explained that during the 1960s there was a craze for weird presents. He had bought a stack of gold pans and, as the craze ended, had been left with about a dozen.

I was now 21 years old and had about 11 weeks of glorious summer before I was due to start work for the Forestry Commission. I packed a thin sheet of canvas, a cooking pot and army sleeping bag and hitchhiked to North Wales. The stranger I had met in the mine some weeks before had told me to head for the River Mawddach and work below a massive waterfall. It took me two days to get there and the beauty of the place was mind-blowing. I set up a camouflaged camp on the peninsula where the Avon Gain and Mawddach meet. Another curiosity of this time, almost 40 years ago, was that the forests were relatively empty of people. Mountain biking had not been invented and commercial recreation seemed innocent and in its infancy. Our mountains and forests still seemed wild, the time before well-meaning and enthusiastic officials carved little trails everywhere and suburbanised our last wildernesses by littering them with information signs. I camped on that peninsula for eight days without meeting a soul – unthinkable nowadays. Certainly I was not looking to be found and I rose each morning at about five o'clock, while the foresters still slept, to light a small fire and boil the day's water. I also had a quantity of self-raising flour with me, along with bacon, raisins, sugar and salt. With these I made two small loaves each morning; one sweet, one savoury.

The gold panning progressed slowly. I had sent to America for a book on how to prospect for gold and, while the book had turned out to be small and was written with an infantile cheerfulness, it was all the information I had and I had brought it with me. My very long days of gravel washing aloneness passed quickly and pleasantly. My persistence, and occasional flicking through the cheery American book, had rewarded me with eight specks of bright Welsh gold. On the morning I was due to leave I had let my fire burn on, as I no longer cared about being discovered. Suddenly a young couple in basic walking clothing stumbled down the rocky slope towards me. I must have invited them to have coffee with me with an unnatural enthusiasm as, while rooting about for a spare cup in my pack under the canvas sheet, I heard the woman whisper to the man, "Why is he so friendly?" And then the man's quiet reply, "If you had been alone up here for eight days, you'd be bloody friendly!"

The trek back home again was pleasantly uneventful. I showed my parents the hard-won gold, "Is that it?" they asked with undisguised indifference. I set my first precious pieces of natural gold in a small pyramid of clear resin, as the stranger I had met in the mine had showed me. Eight specks in about nine days, not impressive really, but at least it was a beginning: my beginning.

Others have had completely false starts. Bob Sutherland had like me, looked for agates and other semi-precious stones in the Devonian lavas at Usan and Lunan Bay to the south of Montrose. Bob had then expanded his collection of minerals by going further afield. He was looking for a copper mine in the Leadhills area and was having no success, so he asked if there was any older person around who would know about the mines. He was put in touch with a Jimmy Murray, but he couldn't help him as regards the mine. "I don't know anything about mines laddie but I know about gold." he was told. So saying, he produced a small bottle that was two thirds full of little gold nuggets.

Bob Sutherland - part of his gold collection has been on display in the National Museum of Scotland and samples can be seen on their website.

This was taken early in the week and was put in my fool bag. It looked smooth.

Later in the cottage details started showing. Victoria's bust is on the left photo, the date, a crown and leaves on the right. This was an 1890 silver sixpence bent when I opened the crack. It should have been 4g but was nearer 2½g.

Shana was complaining that the only picture of her was a selfie. I know the feeling. Here she is gearing up to snipe the Gold Burn. The Suisgill was too high.

SPOT THE SALMON

The high water kicked off the salmon run. It resulted in close encounters of a fishy kind.

The dark high peaty water had another consequence — we had to use torches.

Superficially it looks good but I had more on day one in March than the seven days in Nov. Day 5 was Gold Burn only. I helped more than looked myself.

At one stage Leon was urging Shona to suck up the gold. It turned out to be mica in the rock.

Here he is wondering why there is no gold in good cracks → I had removed it all.

We had a good fun week. This comment and cartoon was in the visitors book in the cottage. We could endorse it.

Came for a weeks gold panning on the Suisgill. Perfect weather for the first 5 days. Lots water every day until rain overnight turned the burn into a torrent on Friday - too dangerous. All found a small amount of gold after relentless persistence dressed in chest waders all day - each day. Cottage is warm and relaxing, lots of stars in the night sky and serenaded by the sound of baying stags throughout the night. — A really lovely week !!!

On the last day a large flock of Long Tailed Tits fed in birches near the top gorge. As usual they did not stop in one spot very long.

THE PROSPECTOR
The selfie.

This area near the ford had gold in moss well above water level. It would take a lot of rock moving to get at the best rock areas

A gold diary recalls good memories but also helps find more gold.

Jimmy took Bob down the Elvanfoot road and put him into a burn there. Bob had it in his head that gold was yellow and every pan he washed had this line of yellow round the edge. 'I was taking it out with a knife and putting it in an aspirin bottle.' Bob said. He had his jar nearly full and had showed it to a couple in a caravan nearby. When Jimmy appeared next day he was informed by the couple that "the 'boy' has a jar full of gold". Jimmy walked up the burn and a proud Bob handed him his bottle. To Bob's consternation, Jimmy took the top off and poured the contents into the burn. He had been collecting barites, also known as heavy spar and in that area it has a yellowish stain. It was three days before Bob found his first speck. His gold panning career may have started with a bit of a hiccup, but Bob Sutherland made up for it in years to come, taking quantities of gold measured not in grams, but in kilograms.

In 2011 I came across a couple of men working just above where I had decided to dig on the Crom Allt. To say that both were working, is not strictly correct. One was doing all the digging and panning while the other took his leisure on the bank. The digger was keen to show me what he had, and it was clear from his methodical working and the gold in his vial that he knew what he was doing.

The digger was Jim Anderson and the man of leisure was his brother Peter. He was one of the increasing band of people who realised that on retirement he would have time on his hands and he had always had a hankering to try to find gold. Since they lived in Lanarkshire, the obvious place to go was the Leadhills, Wanlockhead area. There he bumped into Nelson who showed him how to use a pan and said he would keep an eye on what he was doing. After some time Nelson, who was twelve feet away enquired "Getting any Jim?" The response was in the negative.

This went on for some time and each time Jim said that he hadn't found anything. Eventually an exasperated Nelson asked to see what was left in his pan and pointed out that he was just about to throw out a little speck of gold. By the time he realised that he had spent the day tossing out what he was after, it was rather late and he had no further opportunity that year. As a result the sum total for his first year amounted to three specks.

Fact: *Profitable ores can have particles of gold which are too small to be visible to the naked eye or recovered by gravity techniques.*

The site of the collection of huts erected for the 1868 gold rush at Kildonan frequently makes the passing tourist stop and take a closer look. Today there is a clear place name sign, Baile an Or, gaelic for the place of the gold. There is a shelter with an interpretation board and on the site of the huts there are often parked cars, motorhomes and caravans. In fact the passing motorist will sometimes see a panner only yards from the road. Stan Johnston and his wife Margaret were passing down the strath on their annual holiday when their curiosity was aroused. It was Margaret who was more interested but I remember Stan, a colleague of mine, telling me after the summer holidays that there was a burn up north with gold in it and you were allowed to pan.

A year passed and Stan informed me that the family had spent their holiday panning and had found gold. I was keen to see what he had and asked him to show me. Next day he appeared with a vial well-filled with a shiny mineral. I looked at it, shook it several times, but the answer in my brain was still the same. "I'm sorry to have to tell you this, but what you have is mica", I said. Stan's face fell and he went off with little more being said.

Another year passed and Stan informed me that, undaunted, the Johnston family had gone back and spent their holiday on the Kildonan Burn. He volunteered to bring in his finds and this time it was my face which must have been a picture.

Two weeks digging had produced over ten grams of quality pieces. Gold panning was to be Stan's passion and dictated his holiday destination for years to come.

It's not uncommon for a panner to take a young relative or friend on a panning trip and show them the ropes. At nine, my grand-daughter Nicola had gained my third level certificate. The first level stated that she could pan out the gold in a pan of gravel known to contain gold. The second stated that she could dig her own gravel from a hole being worked by others and show gold. The third involved her going off and finding her own spot and coming back with gold. She achieved that within a day.

To find a youngster who is excited by the prospect of finding colour without adult guidance is very unusual. John Greenwood's father had a garage near a stream in Yorkshire. He remembers being inspired to take a biscuit tin lid and try to find the metal at around six or seven years old. The inspiration came from Bugs Bunny and Yosemite Sam. Bugs was involved with looking for gold in at least two cartoons. John did not pursue the hobby in any way for over two decades. When he met Morag Shearer and they became friendly, he told her the story. The following Christmas, Morag produced two pans. Their first trip was to the Leadmining Museum at Wanlockhead and the Wanlock Water. While trying their luck they were approached by an elderly stranger who said they would not find gold there, but he did give them some instruction and advice on the other equipment they would need such as a pump and a bucket. They then went to the gold panning championships and picked up a little more information. There John met Alex West for whom he laboured in return for the learning experience. He then went to Tyndrum where he met Jim and Peter Anderson working out of their little white van. He did some boulder moving for them and when they were leaving he was invited to carry on in the hole. In 2011 when between jobs, he had plenty of time to travel from Crieff to Tyndrum each day and with hard work built up a total of 38 grams in his first year. I was aware of John long before I met him as his 4x4 would wake me each morning at around 7.30 am as he passed my parked campervan at the graveyard at Tyndrum.

If John got the bug early but kept it dormant, Daisy Thurkettle started even younger and kept going. It was hardly surprising that she should be introduced to the pastime as her father is Vince Thurkettle. In Daisy's own words, 'I've panned for gold ever since I was about three years old and could stand in the shallows and not get blown over'. Daisy's story will be told in a later chapter.

John Greenwood on the Crom Allt on the day I found my biggest nugget. It was good to share my success with a friend.

Shona Richards is another keen protagonist whose interest in palaeontology and the fossils of the East Yorkshire coast led to a desire to find other treasure, gold. As a youngster she had seen men in a Perthshire stream. Being curious she had asked what they were up to. Their response had excited her, but it was years later before she was in the position to do a course at the Museum of Leadmining in Wanlockhead. The course was successful and gold was found. She was hooked. She has described it as "the most therapeutic and exciting hobby with its highs and lows." Since then she has thrown herself wholeheartedly into finding gold, despite living well away from good areas. She quickly graduated to sniping and proved to be the keenest of learners when in my company.

Today, many people begin by booking a course. In the first decade of the 21st century, several experienced gold seekers offered courses, including Charlie Smart, Vince Thurkettle, Kit Andrews and Leon Kirk. Some would-be goldpanners have saved money by being taken under the wing of an experienced panner. In 2014 I was instrumental in helping three groups and three separate individuals find their first gold. I realised that as my labouring abilities waned, I derived almost as much pleasure from going back to my teaching roots.

Having said that, there is still nothing like turning up gold in the pan to add to the collection. I say collection, because I have all the gold I have found, other than small samples I have given away for research or experimental purposes.

Having introduced how I and others began this fascinating hobby, I will continue to recount various tales picked up over nearly 50 years of panning, both at home and abroad.

OF COURSE

'You know, gold ain't like stones on the river bed, it don't cry out to be picked up. You gotta know how to recognise it. You gotta know how to tickle it so she'll come out laughing.'

Howard, the old prospector, *Treasure of the Sierra Madre*

I had been exploring the Crom Allt for a few years, basing our van at the By The Way campsite, when the then co-owner, Jim Kinnell, asked if I would run a course. Colinne and I had become very friendly with Jim and his wife Jean and I could hardly refuse. Jim is a joker and likes nothing better than insulting people in a jocular way. Teachers are particularly fair game and years after the Kinnells left Tyndrum I would still get emails which poked fun at pedagogues.

A date was set and I thought nothing more of it other than where I would take a large group and if I had enough equipment. Yet all the time I was aware that Jim was enjoying a secret that he was not prepared to tell me until he was ready. A few trips later, we walked into the reception area and after the initial pleasantries he turned to Jean and said "Will I tell them?" "Tell us what?" I asked. He spun it out for a while before breaking what he thought would be a bombshell. The course was for members of Mensa. We were aware that Jim and Jean were both members of this high IQ club, but it hadn't dawned on us that this would be our clientele. I took the wind out of his sails by responding "No problem" and went on to point out that a high IQ was not going to pose any problems out of the usual other than some might be completely handless.

The Mensa course. A high IQ doesn't necessarily make a competent panner.

The day arrived, a spot had been chosen, a test for gold content made and the group arrived. I was right, most needed a lot of repetitive guidance regarding the increased angle of the pan as the gravels disappeared. One gave up after five minutes and spent the rest of the day lying on the bank, having bought wellingtons specifically for the occasion. Most got a few flakes to take home, but two showed a real interest and extra effort. They wanted a second day to learn more. One, Andy Farrell, picked up the prize of a pan for the best beginner. Next day I took them to a new spot where I knew there was gold in the cracks in the bedrock. The current was fast and when we disturbed any gold it would dance across the bedrock until it came to a sump or crack where the reduced current allowed it to settle. The delight was to watch this staccato movement through our shuftyscopes then suck up the flakes with a turkey baster. This was a crude form of sniping, made difficult by the need to hold the viewing box with one hand. We had a good day and it was only then that the second member, Angus Ross, a onetime Conservative candidate for Easter Ross, admitted he had in fact done some panning before. Andy was hooked and when his work allowed him to travel north he was to be found in the Crom Allt trying to pan enough for a ring for his fiancée. Years after the Kinnells sold the campsite, a photograph of Colinne panning near where we were working was still on the By The Way website.

By 2011 I had got to know Leon Kirk so well that he asked me to be part of his team when he took a stand at the Galloway Country Fair, held in August at Drumlanrig Castle. Colinne and I turned up the night before and I helped John Hooper and Davie McMichael with the setting up of two of the BGA's troughs. Next day Jamie Shepherd and I did a steady trade helping paying fairgoers (mainly youngsters) to find the three flakes of genuine gold we popped from a vial into a pan with a cupful of builder's sand and gravel mix. I don't remember any going home without their three bits. They got a good deal for one pound considering the effort of finding the gold, setting up the troughs and the expertise provided. For two days, Jamie and I were kept till it was too late to get a beer in the beer tent. Colinne in the meantime was becoming a trusted helper on the stall. When it came to 2012, the cost of a pitch had risen so much, Leon decided it wasn't worth the effort.

The problem with running gold courses is that you introduce more and more people to look for what is a finite resource. There comes a time when an area becomes so well worked that it is very difficult to find enough to send the client home with a satisfactory sample. While pumping gravel for a Leon Kirk course in the Mennock one summer, I was aware that we were being closely watched. Later in the Hopetoun Arms in Wanlockhead I got talking to the same two gents and it transpired they also had been on a course, with another provider. Leon's clients were going home with a very visible line of gold in a vial, while the two spectators, on showing me their vial, had at best, several specks.

Leon was far from the first to use the Mennock Water for panning courses. Charlie Smart was ideally placed for running courses there, as the burn ran past the bottom of his garden. He set up Scotland's first goldpanning school. He thoroughly enjoyed it but what had started out as a hobby was proving so popular that it was becoming a business.

Fact: Gold medals are traditionally awarded to winners. Olympic gold medals in 1912 were made of solid gold; currently they are covered in six grams of it.

"I didn't retire to start a whole new career", Charlie is reported as saying. He taught the guides in the Museum of Leadmining to pan and then handed over the business to them. They showed people how to pan in the troughs at the museum, but were not running practical stream-based courses. Leon's courses were also proving popular, but rather than run them for personal profit, he teamed up with the museum and wound down his own courses. By 2016 the museum's bookings were regularly into double figures and Leon called me down to help cope with the numbers. I would arrive the evening before and pan enough to demonstrate panning technique and if necessary add to the next day's take. This suited Leon, as he preferred to demonstrate the use of a sluice. I could sense a certain tension in Leon at the start of every course, but gradually he began to relax as we had more and more successful days. He nonetheless was always pleased when I mentioned that there was a certain buzz from the clients at the end of the day. This took the form of positive verbal comments, handshakes, thumbs-up from the departing cars and best of all, a spontaneous kiss and hug from some of the ladies.

The museum asked for feedback and there was one comment that made me change how I operated. Our 'gravel monkeys' (I don't know where the term came from) John Greenwood, Carlo Andreucetti and Danny (one per course) worked hard to provide gravel with heavies and hopefully gold. We had worked that spot without moving as it provided adequate close-by parking, a safe descent to the burn, a calm shallow pool for panning and a steeper area for the sluice. There was also rock nearby to show likely spots for sniping and crevicing.

Inevitably the gold did not appear in every bucket and therefore every pan. It had escaped me that this individual had not seen gold appear in their pan. As a result, on subsequent courses, anyone who had not found gold from the buckets of gravel in their first pan had two or three flakes added to their second pan under their eyes. It had the peculiar effect of making them very nervous and even more reluctant to wash out any gravel.

In gold panning competitions, the need to provide a large number of roughly uniform flakes can cause the organisers headaches. Natural gold for competitions can be bought, but around 2012 The British Goldpanning Association (BGA) used carefully cut pieces of processed gold. This idea seemed to have been taken a stage further by one course provider. I was shown a vial by a beginner after doing a course and it definitely contained bits of cut-up jewellery. This of course will behave like raw gold and using it does show panning technique. What it does not do is give the satisfaction of being able to say that the gold is from such and such a stream. It is also not in the raw natural form which has been released from the local rock then subsequently beaten in the gravel. At the end of a course, if time allows, Leon will often produce a small LED hand held-microscope. He invites each individual to view the pieces in a glass vial from below. This inevitably elicits an "Oh wow"! Yet another aspect of this material is revealed to the beginner, its rough surface beauty.

The difficulty of being certain of finding gold for a course nonetheless remains a worry. It is essential that the instructor knows the spot well or has arrived in advance and tested to make sure there is gold. Gold taken from that stream can always be salted into buckets or pans as a last resort, but it should be from that stream. Latterly we favoured this method but uneven salting could give one pan with numerous pieces and another with nothing.

Carlo Andreucetti as 'gravel monkey'. What happens if the gravel monkey does not strike colour? This is when a pre-dug vial of gold has to be used.

On a couple of occasions in the Tyndrum area, I had volunteered to dig for Leon at spots where I had found some gold but had not been there for years. At both I found some straight away, but the supply quickly ran out and I was getting empty pans. My job in these days was the gravel monkey and Jamie Shepherd was demonstrating the panning. I had to make sure there was gold in each of the participant's pans. At both locations we were close to bedrock where the role of bedrock morphology in the deposition of gold could be explained graphically, but on the second course, we had to end up going to a gravel bar in an adjacent stream where we did dig a steady supply of fine flood gold, but the rock element was missed out. Later Leon confided that although the clients had gone away with good samples and a clear understanding of how to extract gold, they had missed out on what he felt was one of the essential elements of his normal course.

At the other course, a change of direction of our digging saved the day and took us onto a gold layer just above bedrock. Early on, on the first day of two days of courses at the Pinetrees in Tyndrum, I was aware of a young lady hanging around where I was digging. I assumed that she had come to do the course and pointed out where Leon's van was. It was only later when she appeared laden with gear and began digging just above, that I realised she was not one of the clients and was doing her own thing. She had a sluice, a very sophisticated grader of American origin more suited to scientific gravel sampling than the quick sorting of gravel from large stones. There were also several pans. She dug all day, outlasting a group who had to catch the ferry to get back to Ardnamurchan. I was aware she had been speaking to Leon. Next day I discovered she had invited Leon and Jamie Shepherd for a curry in her caravan.

Learning from an expert is the best way!
Leon Kirk shows George Marshall how to snipe. They have cleared down to bedrock.

Day two had only three clients, one of whom had car problems and delayed the start. I was on top of gold provision and at lunchtime had time to look at another spot upstream. The lady, who I discovered was called Chris, was digging in the same spot, watched by her non-panning friend. I came back and enquired how she was doing. "I haven't found anything." was the response. I was gobsmacked. A day and a half of steady digging had passed without my noticing she had not used her snuffer bottle. I invited her to try the spot upstream and immediately realised her panning technique left a lot to be desired. She was not removing enough gravel from the surface to get down to the heavies with any speed. It took a couple of pan-loads before she had the confidence to tilt the pan progressively steeper. Eventually she uncovered a few small flakes, but like so many beginners was looking for the gold in the wrong place in the pan. That rectified, she was left to her own devices. The course ended and I told Leon that 'Curry Chris' as I had named her, (to distinguish in future dialogue from other Chris's in gold panning.) was still pumping further upstream. We approached quietly and at distance noticed the unmistakeable action of a panner sucking up gold with a snuffer bottle.

There is not a great deal for kids to do in Tyndrum, especially in the summer holidays. Effie Crompton, a local resident, ran a week's activities in the summer for what were called the 'Strathfillan Saplings'. Would I show them how to pan? I had made Effie and her husband, Professor David Crompton's acquaintance by showing them what I was finding as they walked their dog along part of the West Highland Way. As with most other local residents, they knew full well that the Crom Allt contained gold. This was not always the case with those from further afield. Many walkers attempting the 96 miles of the West Highland Way would ask what we were doing. Some thought we were fishing but on explaining and even showing them what was in the pan, it often turned into a photo shoot. From the responses to my curiosity as to their home country, we must be in photo collections on a few continents. There was one memorable response to my "gold" answer to the question, "What're you looking for?" As the group walked on, the broad Glasgow accent of the enquirer said in a dismissive tone, "Aye, that'll be right".

Effie duly arranged a visit to the Cononish mine where Chris Sangster, the general manager of Scotgold at the time, showed us around. The core shed was interesting as were the sulphide ores which stained the walls in the adit. but in many ways the highlight for me was the polished sample of ore from the mine. It had visible gold. After a barbeque near the mine all the saplings successfully panned gold at a safe spot near Tyndrum.

In 2012 I was approached by the new owners of By The Way to run a course for a group from Fettes College who were coming to camp. Their housemaster, David Hall, ran a trip every year called the Arran Challenge after the school house in his charge. For years he had provided a different challenge on every trip. David admitted he had panned very unsuccessfully. If anything, he was more eager to learn than the pupils. Since the group comprised of around thirty, Colinne and I had dug several holes each producing a few specks per pan. Being a school trip, safety was our prime concern and proximity to warmth as important as the quality of the gold gravels. Despite the fact it was June, one very wet youngster had to be taken back to base before hypothermia set in. I had mentioned the night before how gold could bury itself deep into cracks in the rock and one of the helpers in the Fettes party, Tom Hawes asked if I could demonstrate this. I took him a short way downstream to a spot where there was fractured bedrock. I spotted a loose bit of rock, prised it off and cleaned out the sediment in the hope there would be a speck or two. He panned it out and I could see early in the process that there was gold. His eyes were on stalks as about 0.1g of small flakes appeared at the end. I provided each person with a starter piece, but they all went home initially with gold of their own finding. This was despite the best efforts of the Tyndrum midge population. I wrote 'initially' as there was a report of a broken vial next morning.

The likelihood that many if any of the youngsters in these courses take up panning is very low, though those paying for a course are far more likely to continue. Most who do a paid course listen and work hard, yet I have been around when a client has spent most of the day sitting on the bank and chatting; not

The author demonstrates the sluice in action to Arran House pupils from Fettes College.

Getting a stretch of water with the correct gradient and current speed for a sluice can take time and sometimes quite a bit of stone moving effort (see photo on page 26).

the attitude for a successful future in gold panning. As it is the more expensive way to start gold panning, a course should guarantee you find your first gold, but thereafter you have to put in a lot of time and physically demanding work to put what you have learned into successful practice.

At the Fettes course I was aware that David had been soaking up the information given. Two years later he appeared at Leon's stall during the gold panning championships. Colinne and I were on sales duty. Beaming, he proudly lifted his wife's hand on which was a gold band. He had used what he had learned to pan enough to have made, that most prized of items, a Scottish gold ring.

Requests come from unlikely sources and not always directly. My bi-annual dental check-up was about to begin when my lady dentist started the 'put you at ease conversation' with a query as to how easy it would be for her grandchildren to find gold. Like so many grandparents these days there are times when she was in charge and looked for things to amuse, usually in holiday time. I mentioned places, but said that without tuition it was unlikely they would find any.

A few days later I phoned the surgery and explained to the receptionist that if Mrs H. was interested, I could arrange to meet her at a suitable burn. It was not long before the phone rang and a trip was planned. I had added that I could guarantee that gold would be found. On the evening of arrival I discovered there was a grandson and a granddaughter, the latter being very young and expressing absolutely no interest in the proposed activity. Her brother on the other hand was excited by the prospect. That evening I demonstrated the properties of gold in my van. I also gave a short lesson on plate tectonics using two Tunnocks teacakes (chocolate marshmallows) to represent the earth with its crust. Fracturing the chocolate crust allowed the marshmallow mantle to ooze through in places. You get the idea. The two hemispheres were scoffed immediately after.

The spot chosen proved to have gold near the surface and Kieran was producing several flakes per pan from my pumping. His sister was in the meantime thoroughly enjoying herself further upstream trying to knock down a tower of stones perched on a mid-stream boulder by throwing stones from the bank (as a kid this was called 'dolly molly shots' in our house). I left Kieran to his own devices to speak to his mother and granny on the opposite bank. He came across beaming and showed a pan with seven bits in it. Level two, I thought, i.e. he has dug and panned gold from a spot known to have gold.

It was lunchtime and after eating, Kieran said 'Thank you Alan'. 'Does that mean you have had enough?' said his perceptive granny. 'Yes' was the response. We packed up and headed back to the vans. Had he persisted he would have tripled his take as I discovered next trip. Nonetheless he had a small vial with a complete line of gold when levelled. Granny wanted to put it somewhere safe, but he insisted on holding on to it. We wondered how long it would survive.

A few weeks later I had an email from Mrs. H. to say Kieran had given a presentation in school on what he had done in his holidays. It had been voted by the class as the most interesting, his still intact vial being the star attraction.

Requests for expertise are not confined to the younger age range. In 2014 I received a request from my local B.A.L.L. (Be Active Life Long) group to do a talk that I had already given to the local WI. Since there was an overlap in membership I said I would do something other than alpine botany and rockeries but would not tell them what. The talk was called 'Journey to Recovery'.

I could see their faces fall as they thought I was going to bore them with my medical history. I did nothing to disillusion them when I put up a slide of symptoms, including restlessness and needing to spend a penny (a lot more than that in reality) but the pun had them fooled. There was also the nightmares (where I was fruitlessly digging holes) and paranoia (I was being followed and watched). The final symptom was 'seeing gold spots'. This at least gave me some respite I informed them. The third slide had two words, 'gold fever'.

At this point they realised they had been misled. A brief talk on recovery methods and worldwide journeys looking for gold was followed by a dry panning competition.

The winner got nuggets as a prize, a box of frozen chicken nuggets that is.

I thought this was a one-off, but I had requests from some of the 20 plus other local B.A.L.L. groups, lunchtime clubs and sheltered housing gatherings.

How do you run a gold panning championship in little over a quarter hour in a hall? My solution was to dry river Spey sand and add a given number of gold sprayed lead lumps to each pan. The contestants were given a straw to blow away the dry sand before popping the exposed 'gold' in a vial. At one event I realised I hadn't emphasised the blowing technique clearly enough when one lady started sucking.

Colinne and I were once asked to show some holidaymakers how to pan. We had a hole producing a steady return of fine pieces. The family arrived, a demonstration was given then their pans were filled with dirt. To begin with we helped the adults who were being too cautious. It's a frequent fault of beginners. The water was being taken into the pan and released without the outgoing water carrying off the surface gravel. Intent on rectifying this, neither of us had noticed their son's effort behind. It was the exact opposite. Before we could help, he held up a gravel-less pan and said 'Is this gold?' Colinne looked then showed me the largest piece of gold we had seen in a couple of seasons. What had he thrown away?

I've said to Leon on more than one occasion that the more people that are introduced to the hobby, the less gold there is to go around and the greater likelihood of problems arising. You cannot know beforehand the character of participants. Almost all are very enjoyable to work with and really appreciative of your efforts. In late 2015 I began to detect a change in Leon's attitude to running courses after one would-be panner asked on the phone how much he could make per day and another was causing environmental problems on one of our regular burns. It was about then that Leon decided to channel his teaching efforts through the museum as this would provide a financial help to the cash-strapped tourist attraction.

The rise in the demand for courses was undoubtedly fuelled by the popularity of TV programmes such as *Yukon Gold* and *Gold Rush Alaska*. However, there was an attempt by a TV company to run a similar programme in Britain. At one of my first championships I was aware of negotiations going on to assemble a team. Leon spoke to the Queensberry estate about getting permission to do it on one of their burns, though other areas were under consideration. Eventually the idea died a death. I had been keen to be involved, but on reflection decided it was a bad idea as these programmes have a habit of trying to set participants against one another. It could have spoiled good friendships.

In fact we did make a film for a French TV company in which Leon was shown as the instructor but his beginner pupils consisted of seasoned panners Jamie Shepherd, Colinne and myself along with relative newcomers, Gareth Moore, Emil Heinneman and his partner Louisa. The filming was done in one day and came out very well, with no contentious issues as a result.

Nowadays, the ability to make a film and post it on the internet is relatively easy. Some are worth watching, some are not, and some set out to mislead. One of the latter gave hardly concealed clues as to the general area then showed the panner lifting numerous nuggets from surface gravel. Leon contacted me about it and after a little geographical fieldwork on the internet, I was able to send him the grid reference of the spot and assure him that it wasn't a Bonanza Creek.

Although it is easy to make a film these days, there is still a place for the photograph. In April 2017 we were joined by the photojournalist Paul Jacobs who proceeded to take around 2,000 shots. Some were used by the press in the build-up to the World Goldpanning Championships in Moffat later in the year. After the course I took Paul to another stream and donned my drysuit as he wanted shots of sniping. After a couple of minutes I was able to show him a good flake in a crack. Not only did he shoot from above, but he also took the gold flake's view of the snorkelled sniper. I was looking in a shop window prior to the championships when I saw myself as my quarry sees me on a newspaper page.

The sniper at work. This is a very relaxing if sometimes hypothermia inducing method of looking for gold.

At every course we ask what has made the participants sign up. In more cases than not, it is the excitement generated by watching programmes about gold on TV. Leon assures me that most won't do it again, but I think many get the fever, if they haven't already arrived with it. Judging from the number of people who admit to watching *Gold Rush*, perhaps every established hobby panner should be grateful that not every viewer has had a stirring to go on a course then continue to find gold for his or herself.

Fact: Crustal gold probably came from meteorites hitting Earth over 3.8 billion years ago!

Leon insisted that Colinne wore one of his hats for the filming.

Leon introducing a Museum of Lead Mining course on the Mennock. We did warn them that if they took it up as a hobby, and some do, it might take them long enough to find their own gold.

A likely looking spot in the Buccleuch Estates permit area. It had some good sniping territory with the odd picker.

PANS, PUMPS and TURKEY BASTERS

'To invent, you need a good imagination and a pile of junk.'
Thomas A. Edison, 1847–1931

There are so many different types of stream. Each gives rise to slightly different situations in which gold can be trapped. In some it is freely distributed through gravel, in others the gravel can be devoid of gold, but it is under the gravel in rock cracks. Having looked for gold in hundreds of streams I have not infrequently wished for a tool or implement to resolve a situation where I was not making as good progress as I would have liked. Often, after the trip was over, I started sketching different designs for an implement that might do the job next time. It is one of the joys of this pastime. It stimulates the inventive mind and I know I'm not the only one who enjoys this aspect of gold panning.

When I realised I was totally hooked, I wanted a proper pan. I went to the local ironmonger in Banff and to my astonishment they produced a metal gold pan. It was 45 centimetres across, 10 centimetres deep, had a dome in the base and a slight dip then a bulge just below the rim. I haven't come across a domed base since, but it moves the gold to the edge of the base and then with tilting holds it at the far side. I used this pan until the mid nineties when I started bringing home plastic pans from Colorado. Metal pans are still requested. I've seen a photo of Vince Thirkettle using one and Davie McMichael from Thornhill has a substantial collection of them. I knew he would like mine and when I took it to a course for display, I was proved to be correct. Most of the gold panning tourist spots in the USA have at least one cabin with rusting pans hanging on a wall, but there are still fans.

In 2014 I phoned Leon one night and he had metal pans in the oven. On a later call, I was informed that they were in the bath. This process of blueing or seasoning the pan is done to remove manufacturer's oils which might make fine gold float out.

It also inhibits the rusting process. In goldrush days this was done by putting the pan on an open fire before cooling it in a stream. Modern plastic pans also come with oil on them. New ones should be washed in wash-up liquid and rubbed down with sand and gravel or a fine emery paper before use.

Other than breaking the handle off an aluminium frying pan in the early days, I have never thought of trying to make the basic essential of gold seeking, a pan. There are so many pans on the market that I don't think I could match, far less improve on my favourite pan, the Garrett Super Sluice.

In 2011 I did meet a man who had made a pan and a very successful championship winning pan to boot. At the end of the 2011 Scottish and British Gold Panning Championships I decided to have a very quick look at the Wanlock Water at Spango Bridge to see if the gold had travelled that far. Having found a few specks, I headed through Crawfordjohn and up the B7078 and parked in Happendon Services. I was eating a sandwich when I recognised the couple walking past our van. I jumped out and said hello to Erich and Ilse Baron, an elderly Austrian pair who had been competing at the championships. Their travelling companion and his wife had agreed to meet them at Hamilton services. I quickly realized the similarity between Hamilton and Happendon had caused confusion.

David McMichael with Two Dogs Colin (there is a One Dog Colin).

Erich and Ilse Baron with a rarity, a very efficient homemade pan.

Erich had particular difficulty in pronouncing the two names. A quick mobile call ascertained that their friends were indeed waiting for them further up the M74. Next the conversation turned to a more immediate problem. They had run out of cooking gas for their van and I was able to direct them to a likely source. Ilse had won the Scottish Open competition and was kicking herself for losing bits in the British, when her normally impeccable style failed her and she missed out on the medals.

With the gas issue out of the way, I congratulated her on her success, complimented her on her smooth technique and said I found her speed pan interesting. I'd noticed from a distance that it had deeper grooves than many of the others. It turned out to be far more interesting than I had initially expected. Erich, a former World Veterans Champion, rushed back to their van and appeared with the pan. In German with the odd word of English he proceeded to tell me how he had made it. On close inspection I realised it was not as regularly grooved as the fibreglass and lathe-turned plastic pans. These are often referred to as Ferraris because of the speed with which they can reveal competition gold. He turned and pointed towards the trees round the car park and with gesticulations got me to realize that it was wooden. Colinne, who had returned from the services area, added the word *baum* and this was confirmed. He kept on pointing to a birch tree and on showing him a picture in my wild flower book, he confirmed "Ja, ja Birkenbaum". The next bit of information was the dimensions of the block used. Fifty centimetres by fifty was emphasized and Ilse pointed out that any larger would not be accepted in competition. He then demonstrated how he made concentric circles and cut them out starting with the inner one. This was to allow undercutting. The outer riffles were less deep and the central gold catching area was cut away and a base with a central knob added. The whole thing was then lacquered in black. There was much hilarity as he mimicked the heady effects of using aircraft dope. I pointed out our birch trees were too narrow to cut a block. He immediately realised I hadn't appreciated it was plywood. Holding the edge in the strong sun revealed a faint hint of the layers. Somehow he conveyed that it could be purchased in a hobby shop. The whole process was repeated at least twice and he encouraged me to make one for myself.

The conversation then turned to the rest of their holiday. Ilse was able to tell us that they were heading for Oban. As gold panners love to get foreign samples it was only courtesy to point out where their journey would cross the Crom Allt at Tyndrum. We wished one another a safe journey and I said I hoped I would see them at next year's championship, but Ilse said that they were so old now that they did not plan that far ahead and took life one day at a time. Much to my disappointment, they did not appear the following year. However, Moffat World Championships in 2017 did tempt Erich and Ilse back to compete. Erich died just over a year later aged 86.

Our first plastic pan was the Garret Gravity Trap. We had two until one floated away un-noticed while speaking to a spectator at Baile an Or on a windy day. Despite a thorough search of the peaty water, we never saw it again and given its robust nature it is probably in the Moray Firth. The pan has a limited base width and was not the best for

A selection of my pans. I favour the robust Garrett Supersluice, top right - I have three. Despite constant use they are wearing out less quickly than me.

processing large amounts of gravel quickly. Nonetheless I sometimes carry it as a small prospecting pan. I brought back several wider pans from the States several of which suffered from lack of rigidity and or poor riffles. There was one pan on sale in Britain with bars protruding from a shallow sloping side. If gold got on the rim side of a riffle, it could not be dropped back to the base without an oscillating motion which was very difficult to master.

> *The pan which quickly emerged as my favourite was the Garrett Super Sluice.*

Despite having only two riffles, its robustness means it can survive lumps of clay covered rock being tossed in. The width of the base and angle of side allows gravel to be moved vigorously to allow the gold to drop to the base. I've seen another well-known make of pan broken, by the sort of activity I subject my Super Sluice to. Colinne prefers the lighter Klondike which has more riffles. It is also the pan used in the knockout competition at the home gold panning championships.

There is a third style of pan used in the tripan competition at the championships. It is the batia, an open conical shaped gold sorter favoured in many third world areas. I have only come across one person who enjoys using it in a stream.

Davie McMichael took me to a river where he had promised to show me a high placer. We were struggling to find gold high up, but the gravel close to the river produced a few flakes. Davie then cleaned out an open crack in rock and reduced his batia pan-load to a little black sediment at its centre. He knew he had a good pan but exposing and isolating the gold was a slow process and I could see why there are few batias used in Scottish streams.

Everyone has a favourite pan but I would like to have one which has the robustness of the super sluice, the same number of riffles as the Klondike and a wide domed base like my old metal pan. After designing his sluice range Leon Kirk turned his attention to a pan as he felt there was room for improvement. He had taken on board some of my comments, but was adding other refinements. As the tooling for such a venture is very expensive, it was difficult to have a test prototype made. I was convinced that at least one of the features was designed to attract the buyer rather than to be of indispensable use to the seasoned panner. It reminded me of some fishing lures which were more likely to catch the angler than a fish. Perhaps this is not the best analogy, as a well-designed basic pan with additions, will still catch gold. Time will tell, as my conversations with Leon were ongoing at the same time as editing this chapter.

Of all the tools whose introduction has made a great impact on gold recovery in Britain, the gravel pump must surely rate as the foremost. GFS Adamson in his book *At the End of the Rainbow* calls it 'the Kirk/Laird sooker' after his two friends who were developing it. Today it is sometimes referred to as the Henderson pump, after Alf Henderson who showed me one of his originals at the 2012 Gold Panning Championships. The story goes that Alf was taking much more gold than others but hiding the secret weapon which was giving him an edge. They became much sought after and it was not advisable to leave one in view in your car. In the nineties, Goldspear (UK) Ltd were offering pumps. Their colour catalogue had two photos of a pump in use on the front. Inside was a diagram giving instructions in its use, though here it was called a gravel sucker. Today, after a pan, they are the next purchase for many hopefuls.

Alf Henderson and Malcolm Thomas, who was president of the BGA when taken.

Fact: Nuggets are rare, making up less than 2% of all native gold ever found.

My first atempt at making one was an absolute disaster as I used a plastic map case lid screwed onto a broom handle as the suction. Needless to say it did not survive the rigours of its first outing and accounted for no gold whatsoever. I then visited my local plumber and was allowed to go through his collection of leather washers. He furnished me with pipe to fit. An old ski pole with an untapered aluminium shaft (you don't get them like that today) formed the rod. The pump had reasonable suction, but it suffered from a major drawback. Between uses, the leather dried out and it had to be soaked for several hours before going on a trip. If I forgot, it was useless for the first few hours. It was Alan Price the Orcadian panner who showed me his beautifully crafted homemade pump using a tennis ball in a plastic drain pipe. I bought a set of taps and dies and used aluminium rods from DIY shops for the shafts. If not as robust as plastic covered threaded steel it has the advantage of lightness. Eventually I settled on a T head solid aluminium handle which could be screwed on to the shaft.

The demand for good pumps increased in the teens as the hobby, fired by TV programmes, caught the imagination of a greater audience. I can recall two occasions when Chris Paterson turned up at the championships with a bin bag full of them. They were handed on to Leon for sale on his website. He was not the only one making them for Leon. Charlie Smart of Mennock Cottage was another supplier. By 2014 supply was not keeping up with demand. Although I was asked to become a maker at this stage, I had enough on my plate. For a while, Leon had the use of George Nutt's workshop to turn out bits, including the metal protective end. He was constantly researching materials to make the pumps more robust and more uniform, while increasing the suction pressure. Eventually he set up his own well-equipped workshop at his mother's house in Castle Douglas. He would travel from Glasgow to make his 'Super Pump', the demand for which put him under a lot of pressure.

Circular plastic rain pipe is not totally rigid. The suction in a well-made pump can lift a stone which can jam in the pipe and distort or even crack it. The secret of release is to identify the two points which are being pushed out then squeeze the pipe even more at right angles to the line between these spots. This normally releases the touching spots, but in doing this, there is always a danger of cracking the plastic. Many people put a large jubilee clip at the business end to help to keep its shape and to prevent wear to the plastic.

Gravel pumps need about six inches of water otherwise they can draw air. On numerous occasions, having cleared gravel off a near-surface rock shelf, I knew I was losing gold as I could not lift the last of the gravel with the full bore pump.

A former pupil came to my rescue. He was in charge of the workshop of one of my local engineering firms. I described what I wanted and he put an apprentice on to it there and then. I never go without what I call my 'reducer', a conical shaped nozzle with a diameter at the business end of 2 centimetres. It jams into the end of my pump in a matter of seconds and can be removed just as quickly. It works in any depth down to about an inch and is used to suck up the final grains. The number of times it has produced a good flake from a muddy corner when I thought there was nothing left, is legion. Because it is smoothly tapered inside, it rarely blocks, unlike my early version made from copper plumbing pipe. George Nutt tried to persuade Leon to develop a similar end. It also has the added use of producing a high powered jet of water which can be used for 'hydraulicking'. This is the process of blasting gravel and gold from cracks. It is the panner's version of the giant monitors used in goldfields like the Klondike.

On one occasion on the Allt Ghamhnain, I had pointed Colinne in the direction of a very narrow crack in extremely hard rock. She had managed to take a few small flakes, but when I gave it a prolonged blasting with water from the nozzle, a fine picker popped out. On another, I was waiting for a group to assemble on the Crom Allt. On spotting a subsurface crack I was informed Jamie Shepherd had scraped it the previous day. I had nothing else to do till the group assembled so I blasted one end with a jet of water then inspected the results as the muddy cloud settled just beyond the other end. Was I disappointed? Not a bit of it. I kept the resulting pieces in a vial to use when demonstrating the panning technique and it could make some eyes light up. (I later lost this gold when a well-meaning helper washed out my pan at the end of a course before I had sucked it up – I learned a lesson that day)

Another end fitment that some like John Wilkinson (One Tooth John) and John Cathcart (Captain Gadget) use is a drainpipe 112 degree elbow. This prevents the gravel falling out prematurely. I prefer to have the tennis ball right at the end of the tube to get the maximum suction then have my pan or bucket close to the point of extraction to minimize loss. A strong pull combined with a dipping motion of the top of the pump prior to tipping, prevents gravel loss. Nonetheless normal size pumps can choke if the top is pushed below the surface and water gets in between the ball and the top. When panning was allowed in Welsh streams, John Hooper wanted to get far deeper than a normal pump would allow. He made a six foot pump using a double drain rod as the shaft. The angled end allowed him time to do a double stroke pull, lift it and drop the gravel into a pan. At the time of writing I was trying to develop a shorter one draw pump which had a three foot breathing tube straight out of the top to replace the side breathing holes. The problem I hadn't solved at the time was how to get a tight waterproof seal between the shaft and the head without making it difficult to pull.

Brian Meehan with a range of gadgets behind made by John Cathcart, 'Captain Gadget'.

Pumps will inevitably break and it pays to carry a spare one, if not on your pack, at least in the car. Mark Oddy drove all the way from the south coast of England to join three of us on the Suisgill. Within the first hour his pump tennis ball had cracked and he was facing a week without this essential tool. Fortunately I had a spare in my van which was only a few hundred yards away and successful digging continued within the hour. There are rarely fewer than three full sized pumps in the gear storage area of my van. When I say full sized, my wife prefers a much shorter one to me, but I also have ones that sit discreetly inside a rucksack.

In the eighties Stan Johnston heard about Australian bait pumps. Living in one of the Speyside villages with a well-known tackle shop he didn't find it hard to get hold of one. Although quite expensive at the time (over £40) it is the only pump Stan has ever needed and is still going strong. It has adjustable suction and is a little shorter than most pumps today, but it has taken a lot of gold for Stan and Margaret and doesn't owe them a penny.

Sometimes holding a pump can be a strain on the hand and I toyed with several designs of handle. Almost immediately after, I was at the Crom Allt and bumped into Dan Haddow. He had fitted one to his pump and said it was worth doing. I did try one but am not convinced that you have quite the same control. Until I find my hands can't hold the barrel, I'll continue to do without. Some wrap and tape rubber round the holding area to reduce the slippiness but this has to be renewed quite frequently.

I am aware that others have experimented with very wide bore pumps, very long pumps for deep water, narrow ones and ones with a valve on the end which holds several draws worth of gravel before needing the top pulled off and the gravel tipped out of the top end. I even had one with a long glass-bottomed tube attached like a rifle telescopic sight so that I could see exactly where I was drawing the gravel from. It's all good fun if you have the time, but I seem to return to the simplest of designs in the end.

The ability to see what is happening below the surface is of paramount importance. Scottish pearl fishers, when pearl fishing was allowed, used a jug to help spot and identify the most likely mussels to contain a pearl. Often these were made from a berry picker's luggie (bucket) with the bottom replaced by a disc of glass. (Pearl fishers often used work in the Scottish raspberry fields to raise cash to fund trips to the rivers.) I don't know where I heard the term 'shuftyscope', but it is the one I used for my own versions. For a long time, perhaps the best known one available was one made by Malcolm Thomas (a President of the British Goldpanning Association) from black painted large beans cans with a glass insert. Leon named it and sold it under the name 'beanseen'. I made numerous viewers from various plastic drums including a former garden spray, a kettle, a vacuum flask and an aluminium spotlight shade. The latter made a small but lightweight viewer which I use if I'm prospecting distant spots and I want to keep rucksack weight low. I settled eventually for 125mm plastic pipe as my standard choice. I prefer to use glass as it scratches less easily than perspex, but I have broken a couple through carelessness. Glass does not take kindly to being dumped unceremoniously on rock, nor should you carry back rock samples stuffed inside. You can, however, save rucksack space by using it as a sandwich box. The current one, which combines strength, lightness and a robust handle is fashioned from the casing of my dropped Thermos flask. Leon's 'Caledonian Goldspy' is a lot wider. This bigger aperture gives a very good viewing area, but makes it harder to hold in a current.

Jamie Shepherd rarely carries one, but will have a snorkelling mask on his head and will plunge his face in the water when required, no matter the season. It certainly leaves both hands free and is a lot faster than retrieving a viewer from its safe resting place, which is frequently the bank.

Current strength can be a problem even for the hand, never mind the teeth. My original design has a handle large enough to slip a gloved hand under, taking the strain off the fingers. It is a rigid kitchen door handle, but I've also toyed with the idea of a two part velcro strap which would allow adjustment for any size of hand or glove.

Over the years I've seen many others. The standard design is a bucket with the base replaced by perspex. The most unwieldy was a floating box with a wide clear base. It was left in the current on the end of a string attached to a three kilo iron weight. It worked well enough but I would not have liked to have carried it far.

A pump may work well in gravel, but getting at gravel can involve boulder removal and this needs something more substantial. The favourite tool is a crowbar. It was years before I invested in one, but after breaking several shovel handles, I succumbed. Johnny Marr (Bedrock Johnny) combined his digging tool with crowbar capabilities. The blade of his shovel had sides and a back to prevent gravel loss, but the shaft was also reinforced with metal. I'm sure his design was what led to the insistence in the Kildonan rules on a wooden shafted shovel.

Because of their fragility I changed from wooden shafts to fibreglass. I got my first fibreglass shaft in Walmart in Anchorage, Alaska, a hardware store being my first port of call when gearing up to look for gold abroad. More pliable than wood, fibreglass usually warns you of impending disaster by creaking and will even continue to function with major cracks.

In New Zealand I went through the same process, but was not enamoured by the lack of choice. Next day we were working in Louis Creek where very large boulders have to be moved to get at gravel or bedrock. Having levered away a boulder I was struggling to penetrate the gravel. It took a few minutes to work out what was wrong. I had split the blade. It was useless and reduced me to using a back-up trowel. We were handicapped till we reached Murchison and located the hardware store. The choice was no better, they all said 'made in China'. I asked the assistant for something stronger. She had nothing else, but directed me to an agricultural depot a couple of miles down the road. The man listened to my story, laughed and said "You need a New Zealand shovel". There was no comparison. The blade was at least twice as thick with a shaft to match. I usually carry a ski bag to take shovels home, but in this case there was a weight problem and on my second last day I approached a complete stranger in his garage and asked him to cut the shaft so that I could at least take the blade.

Jamie used to call me 'Shovel Man', but of late I've carried it less and less. I do always carry a trowel. It is amazing how much gold you can take trowelling in conjunction with a pump. But even trowels have been altered in my workshop. I realised when prospecting that I was targeting rock cracks, often in schists. The clay layers stuck to the rock face held the gold and had to be scraped off. I therefore wanted a trowel with straight parallel raised sides, a straight tip and a handle raised above the blade. It also helped to be stainless. After various prototypes, I settled on a modified B&Q Verve trowel.

A long handled shovel can be a useful tool in gravels. I dug, Steve Matthews (right) moved boulders, Jamie Shepherd made a channel to keep the water clean and Leon ...er... 'supervised'.

I mentioned earlier that I think carefully what I pack for a specific burn. If a burn has a lot of sterile gravel on the surface, many like a rake. I've come out of several antique shops with long tined rakes which you won't get in a garden centre today. I wasn't happy with any of them and wanted something that took greater amounts away with a raking motion, leaving only the fine gravel. I cut tines in a spade blade and had a blacksmith make me a strengthened shaft holder which was welded at right angles. It is more like a big hoe, but it works well.

If I deemed my rake a success, my stone removers were not. I had discovered a very promising spot on the Crom Allt where a rock band dropped to some depth. The deeper I went, the better the gold, but at depth I was shipping water down my waders trying to clear cobbles. Heavy overnight rain filled the most promising hole I had dug and for various reasons, I never got back to it. Nonetheless, I was determined to make a grab which would lift out stones at depth without me getting soaked. What transpired was like a litter picker made from bits of surplus greenhouse aluminium. It did lift stones in dry trials, but was too slow and unreliable in water. An alternative landing net-like tool was equally useless. The answer was to buy a dry suit.

There are rocks covering gold which require more than a shovel or crowbar to move. The 'flat rock' above the sheepfold on the Kildonan is such a rock. Who knows how it was moved to remove the gold, (if it was), but the area below was reputed to have produced over an ounce. The River Garry has some very large boulders. It also has a little gold, but it is in isolated pockets in fractured rock. One of the best spots I found ran out under a boulder around five feet long. I determined to pull it out of the way. I need a fencing ratchet cable stretcher for work around my garden and I rigged it up with ropes to an even larger boulder and inched the rock aside. I wish I could say it was worth it, but it wasn't.

Jamie Shepherd mentioned he knew a spot on the Crom Allt where he felt there would be good returns if he could get help to remove a large boulder. We rigged up the winching apparatus and started cranking. Unfortunately the anchor rock was looser than the target one and both moved. Yet again this effort was not repaid with good returns.

I know a number of panners who have a 5 ton jack. This is used not to remove the rock but to prise it up enough to remove the gravel below. The most ingenious rock removal set-up I've come across was by an Alaskan panner, Kelly. He worked a claim on his own with a minimum of equipment. To get at bedrock, he had to move hundreds of rocks. He aimed to move a hundred each day before he started looking for gold. Many he could lift, but some were heavy and awkward. On the wooded bank above the stream he had a winch attached to a substantial tree. When he needed it, he rigged up a car battery which had a remote switch. Once attached in a strop, the winch was activated and I stood in awe as the boulder crawled across the bank to the stacking area.

It is not necessary to move boulders to find gold, but once you have spotted colour, it is important to have the right tools to recover it. For removing gold from a pan there is nothing to beat a good snuffer bottle. Snuffer bottles are not the most visible of tools and are easy to misplace or drop in the water. The one I use on a daily basis has high visibility orange tape attached. Others prefer to have it on a lanyard. Snuffer bottles used at the river will invariably suck up minerals other than gold, but can also be used in the cleaning process to separate what you want from what you don't. I have seen film of American gold seekers working cracks using an ordinary snuffer bottle though I feel it is better to have the greater suction of a bulb. Over the years I have built up a collection of bottles, some of them home-made. The plastic tube from a soap dispenser is of the right diameter and robustness. Several plastic bottles have the right 'recovery speed' to give good suction when squeezed and released. Soap manufacturers constantly change their packaging and at the time of writing I was washing my hair with T Gel. A little drilling and ingenuity is required to fit the tube.

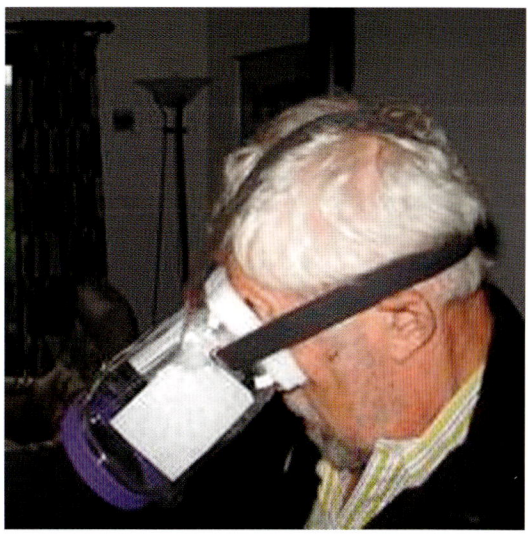

Virtual reality gold panning? My prototype mask for sniping in waders with two hands free, yet not shipping water. Yes, it did hold nuts in a previous life.

As sniping became more popular demand grew for a good bulb with a wider tube than a snuffer bottle. With none on the market, plastic water bottles were adapted to make a larger version of the snuffer. The search for good bulbs was on. I came home with one from that emporium of gold panner's delights, 'Alaska Mining and Diving' in Anchorage. When I tried it in cold water, the bulb stiffened and would barely lift gold.

> *Fact*: For centuries more cerebral effort was expended by alchemists trying to turn base metals into gold using a substance called the philosopher's stone, than was used to make gold mining tools.

Leon was also searching the world for a good sized rubber bulb. After several false trails he got one and made good sniping bulbs. I was also searching eBay. Car horn bulbs were too dear. Many were too small. I did buy one for expressing milk. Since I was a male, I explained to the seller it was going to be put to another use, just in case she thought I was a pervert. It too was too small. Eventually I bought a couple of Sealey battery fluid fillers, removing the plastic tube and replacing it with aluminium; making sure the tube ran well into the bulb to form a non-return trap. Too light to stay static under water, I added a lead collar. This has proved a very satisfactory gold lifter.

If I'm not sniping I may still carry it, as gold can appear on rock in a number of circumstances. It has replaced the turkey baster I used to carry for doing a similar job. An unfortunate accident (described later) resulted in the addition of a jubilee wingspade clip to hold on the bulb, but allow rapid release.

I have never been a great fan of sluices. It's not the sort of experience from which I get most satisfaction. Having said that, I've owned an aluminium Pro-Mack sluice for longer than I can remember. It had a small feed tray, but I asked my brother-in-law who had access to the tools of a school technical department to make me a larger tray. This allowed me to dump entire buckets of gravel at a time and let the current slowly wash away the pile. When Leon produced the 'Hungry Haggis', Colinne saw Jamie use it and was impressed. She promised to buy me one and I was delighted with its light weight and ease of setting up. It immediately flagged up any gold in the gravel. The cleaning process was also very rapid and efficient and if the lower riffles were cleaned first and showed no gold, you knew it had been set up properly. Normally the gold did not travel beyond the third riffle. Compared with releasing a mat, often with great difficulty and washing at least twice, the new sluice was a joy to use. As Leon developed his range of Caledonian sluices, I was involved at various stages offering advice and criticism.

Garnets collecting in sluice riffles. Any gold will be underneath.

An engraving of the gold mining activities on the Kildonan Burn published in the *Illustrated London News* in May 1869 shows various recovery methods. One of them could be a rocker box. As some of the miners were seasoned diggers who had either been to or absorbed the techniques developed in California, Australia and New Zealand, it is very likely that rockers or cradles as they are also called, were in use. R.M.Callender and P.F. Reeson in their *The Scottish Gold Rush of 1869* also believed rockers were used. They made one and tried it successfully in 2005 as part of their detailed examination of the 1869 activities.

At the BGA meeting at Tyndrum in 2015 the guest speaker was Dr. Neil Clarke. Neil had produced the excellent Scottish Gold exhibition in the Hunterian Museum in 2014 and written an accompanying superb book on the subject, *Scottish Gold – Fruit of the Nation*. His talk produced some more information that was not available then. Enhancement of the well-known photograph of Baile an Or during the gold rush had revealed a rocker box outside one of the huts.

In 2013 I was inspired by the plans in *Gold! The Way to Roadside Riches* by Tom Bishop. I set about making a rocker. It was almost finished when doubts entered my mind. "This is heavy and requires at least two people to operate it. Where am I going to be able to use this beast?" The cradle project was eventually scrapped literally and metaphorically.

By far the most ingenious piece of equipment appeared on the Kildonan Burn in August 2010. It was a midge infested day when Dan Haddow revealed his solution to digging gold in clouds of the tiny menace. The Haddow floating tent was revealed to not a little ridicule. Undaunted, Dan headed to a deep area, anchored his tent and began digging. As we walked past, a body popped out momentarily and mocked the "poor suckers" who were going to have a less than pleasant day being food for the midge. Did it work? Yes it did, but I have not seen it since.

I am not a great fan of carrying every bit of gear I have to the water's edge. I think carefully about what I pack for a particular river and if it can't all be carried in one journey, something has to be left. This is where sharing tools with a partner can give you a greater range, but I have seen individuals make two trips from car to river. If they are close, this is no great problem.

Age also has an effect on what you can carry and although most people pack a rucksack, Kit Andrews described an invention of his in a BGA newsletter under the heading "Gold panning gear carrying device". His article went as follows.

"One of the best bits of equipment for carrying all your gear miles up the glens from your parked car, is the electric golf trolley, not the people carrying buggy, but the bag carrying trolley. Discard the golf bag and replace with a lightweight frame and a large garden plastic trug at the base. Your pan, classifier, sluice, wrecking bar, gravel pump, dry suit, coffee flask and sarnies and whatever else you wish to carry up the glen will fit very easily. The motor that drives the spiky wheels is around 200-300 watts driven by a battery of around 26 amps.

To buy new, these are anything from £150 up to a grand, but you can source damn good second hand ones from your local golf club for around the £50 mark. Well-known internet auction sites have them on sale for similar money, but often they are minus the battery, or at best knackered batteries. A new battery for the trolley is similar to your car battery of around £60.

Dan Haddow and his midge-free floating tent. He had only popped out to mock the midge-eaten passing group.

The beauty of this trolley is that on a single charge the trolley will transport all your equipment of 50lb + (who the hell carries 50lb + of gear?) up 1:3 hills, uphill and down dale for approximately 10-12miles.

What I'm trying to say is that for those of us who get winded climbing the stairs to bed, this trolley is invaluable, especially when you try to climb the Kildonan hill from the car park with your rucksack choco-a-block on your back.

Another plus for my gold trolley is that it collapses down flat in three sections, to look like a mini go-cart".

In recent years I have thought of buying an electric mountain bike to keep up the range of my prospecting. Kit's idea would take the strain out of carrying the rucksack on uphill tracks, but it would not get the gear anywhere near the water of most of the burns I have frequented in the last few years. In 2015 I saw a variation of this idea when I passed Nick Eade, the Cornish gold seeker, working a hole in the Kildonan. His tools were carried in a modern golf bag.

I've already described how many panners have an interest which goes beyond gold alone. Many reports have highlighted how gold can be in microscopic quantities in other minerals or rocks. Kit Andrews was known to crush red sandstone to get small quantities. George Nutt had a large mortar and pestle which he maintained produced gold from rocks gathered in the Wanlockhead area. He gave it to Shona Richards when his engineering skills were used to make a mechanical crusher. It was demonstrated at a BGA meeting in late 2015.

Within seconds, the Wigwam's road metal was reduced to dust and quartz samples from the nearby streams suffered a similar fate. Mark Gregory took a pan-load to the water's edge and panned it, but found nothing. Nonetheless there are bound to be rocks and minerals with gold content that could give the gold seeker some fun and a little payback. Leon had hoped George would produce more for sale, but they did not appear. Leon eventually turned to another engineer friend, Tony Taylor to make him one. The end product was a superb looking device and it was duly installed in a friend's gold shed. At the same time various forms of small electronic gold detector were entering the market, but I know of no panner who has the skills to make their own.

Making or adapting tools to help with gold recovery is part of the fun of this pastime and these examples prove the old adage, necessity is the mother of invention, is just as applicable on the goldfields.

Fact: *An ounce of gold can be stretched into a wire over 50 miles long, and beaten into a sheet of over 100 square feet!*

George Nutt and the rock crusher he designed and made.

HEALTH and SAFETY

'My only solution for the problem of habitual accidents is for everybody to stay in bed all day. Even then, there is always the chance that you will fall out of bed.'

Robert Benchly (1889-1945)

We had spent the night in Quartz Lodge in Reefton, South Island New Zealand. After a fantastic breakfast we departed in pouring rain. It soon cleared. The first port of call was the Miners' Hut where the Bearded Miners hung out, talking about gold to any passing tourist who showed the slightest interest. I was invited to try my hand at panning, but when I produced a vial with some good flakes they realised I wasn't the normal tourist and the conversation immediately changed direction. There was a big nugget in town and I was told the owner was in one of the cafés along the street. After a tour up and down we went into a café, but could not detect the excited buzz we expected if a big nugget was being shown. After a coffee and further enquiry we were directed to another café across the street. The group around the table at the door were in animated conversation. I waited for a slight lull then said, 'I believe you have something interesting to show'. I quickly introduced myself and stated my gold panning mission. One of the party, Tony, put his hand in his pocket and produced a nugget that would tip the scale in ounces rather than grams. Almost as fascinating as the lump of gold, was one of the fingers which proffered it. It was covered in a thick and bloody bandage. It had succumbed to the panner's constant threat. He'd dropped a boulder on it, splitting it right open. Surprisingly, I have only a vague recollection of the nugget, yet I can picture the finger clearly. The nugget was, much to my disappointment, not locally sourced, but purchased from Australia. Who has not at some stage or other dropped a stone on hand or toe or even wedged a finger against the rock with a shovel or crowbar? It's one of the minor, but painful risks of panning.

On cold damp mornings I'm reminded of my own self-inflicted gold panning hand injury. If I grip something too tightly, the third finger of my left hand locks in the bent position. It's commonly called trigger finger. I was panning a good spot on a patchy river in the Tay basin and the returns were getting relatively better, though the gold-bearing gravel was proving hard to get. I had spent quite some time filling the pan to capacity and was carrying it to my panning spot when I stumbled and my right hand slipped off the rim, leaving the left to take all the strain. The brain works feverishly at such moments and processes all sorts of options almost instantly.

'Don't let it drop, you've spent ages filling what should be a good pan. If you drop this you're likely to lose a good pan-load in a spot where it will take you hours to recover it, you haven't got hours, therefore don't let it tip…decision made'. I held on as I tried to recover my balance, but the weight was too great and my fingers got bent back as I refused to drop it completely. Bending rapidly I splashed the pan down in the water and grabbed with my right hand, saving almost all the gravel. One of the things about panning in cold burns is that the water is at a temperature that helps prevent bruising. The finger hurt but I kept on panning. Next day I had a strange, finger-in-the-air grasp of the steering wheel and it was weeks before I could wrap my fingers comfortably round it again. Only later did it start locking. Was the pan-load worth it? I seem to remember it was reasonable, but nothing to dance about.

Since earliest times I have had a fear that while developing a deep hole on my own, a large boulder would roll in and pin me to the bottom. On numerous occasions boulders have rolled down, but they have done it in slow motion and I have been able to avoid them. I admitted this fear at a small gathering and was told the following story by John Wilkinson aka One Tooth John. (John now has no teeth. He removed the last troublesome one with the help of a pair of pliers and not a little whisky.)

John had been well up the Glenclach Burn and had developed a large hole below a boulder. He was aware of the precarious hold it had, but convinced his first gold would come from below it, he kept digging. Suddenly, around two o'clock, it fell into the hole pinning his foot to the bottom.

Naturally he tried to pull it out, but it was wedged fast. He tried moving the boulder, but couldn't budge it enough to release his foot and the rocking seemed to allow more material to build around, making for an even tighter grip. Since he had a wetsuit on, he lay back and tried to use the free foot to push the rock off, to no effect. He had a crowbar with him, but it was out of reach on the bank. Each time he struggled he got nowhere and felt a little weaker. An hour passed, then two. By this time, despite the wetsuit, he was chilling and still he had no idea how he was going to get out, if at all. Was he going to have to spend a night pinned in the water? Would he survive the night? Would anyone else be working this burn?

> *Fact*: In 1898 an avalanche on the Chilkoot Pass killed 70 men trying to reach the Klondike.

Fortunately for John, Newcastle Nelson was working above and three hours after John had got into trouble, he came down the burn and was able to release him. My trapped-leg nightmare scenario involves heavy rainfall and rising water. John was lucky. It was dry and the burn level was static.

In the eighties I was panning a spot on the Kildonan at the upper end of the 'Golden Half Mile', but just below the right angle bend, downstream from the sheepfold. It was at a point where the rock walls narrowed and I had climbed down the western rock face into a little recess where I was slowly working down through some reasonably productive gravel. It had been raining most of the day, but clad in rubber chest waders and a green plastic jacket with hood, I was oblivious to the weather conditions. It was at a time when I still used a large metal pan, had no pump or snuffer bottle and delighted in watching bits of gold drop off my fingertip into my whisky miniature bottle with the black tape on one side. In mid afternoon I suddenly was aware of an increase in noise just upstream and looking up saw a wall of dirty foaming water about thirty yards from me and approaching fast. As so often happens, my gear was scattered on the shelf of rock beside me and I realised it was going to be washed away in seconds. I threw my pan up the rock face hoping it would clear the top and not get snagged in the whins and come bouncing down. It stayed there and I grabbed my rucksack and digging tools and snagged them in the whins above my head. The wall of water was already past me and the burn had risen nearly two feet in an instant and was continuing to rise as I climbed the twelve feet to the level above, retrieving gear from the vegetation as I went. Panning was over and I was on the non path side and had an awkward walk back to the road and the bridge at Baile an Or.

On return home I described the event to Stan Johnston. I was not convinced he believed me at the time. Many years later he experienced the same type of event. It seems that when the slopes of the valley of the Kildonan become saturated and heavy rain continues, there comes a point when the rainfall runs off as though the surface is impervious. It reaches the burn very quickly in the upper reaches and rushes downstream as a wall of water.

I told this story in Mike Jones's presence and a year later, after a visit to Switzerland he recounted a similar event, though the cause was a little different. Disentis is in the canton of Grisons and is sited on the north bank of the upper reaches of the Rhine. It is a popular recreational gold panning area with a 20 metre wide river littered with very large boulders. Its gold panning claim to fame is that until October 2011 it was where Switzerland's largest nugget had been found. The campsite had plenty of information on finding gold, but Mike is one of Britain's most experienced panners so was too excited to spend time trying to translate the German.

The water was low and Mike was operating a pan and a sluice having developed a hole to bedrock in the middle of the river in a spot about 300 metres upstream from the campsite. No one else was around. Working in inches of water with his head down, he was suddenly aware that the current had increased a little. Looking up he saw a flood rushing towards him. He grabbed all his gear and headed for the bank managing to reach it before the water went up almost two feet. There was only one problem. He had headed for the opposite bank and there is no bridge in the vicinity. In fact on the side he was on, there is a cliff-like bank which proved to be beyond his rock climbing abilities. He was stuck there until the water went down unless he could cross back. He wasn't going to risk the crossing unaided so he felled a small tree using stone-age technology, a rock. He now had a support staff, but having ventured in with this, he did not relish the complete crossing in a current which would have swept him away if he had stumbled. He was stuck and his provisions were limited. Five hours later the water dropped enough to allow him to escape. Why had it risen so rapidly on a fine Swiss day? The notices in German were warning that water is flushed from a dam upstream from time to time. The locals knew it was coming. Little wonder Mike was on his own that day.

> *Tip from the top*: Some of the best gold spots are in ravines. This is in part due to the fact that they are less accessible and therefore less worked.

In Alaska, while driving south towards Seward, I saw a man with a lot of equipment in a lay-by. I immediately swung in to see if I could pick up any more local knowledge. Metres of rope and other climbing gear were arrayed around, but it was the pans that alerted me to his mission. On approach I noticed he had a holster with a six shot revolver. "That's for bears" he assured me. Looking over the roadside barrier I could see the need for climbing gear as the slope disappeared in a convex curve to the valley bottom and a river which could not be seen from above. On reflection it might also have been to hold a dredge in place.

I have seen fixed ropes set up to help descent to the Crom Allt by a group working just above the waterfall beside the waterworks. As yet I have not resorted to abseiling into any burn, but have had a few uncontrolled descents into the Crom Allt over wet grass, moss and mud. To date, I have arrived at water level unscathed, if rather muddied and a little shaken. In later years I have resorted to taking walking poles to help with descents.

The scariest descent I ever made was over in seconds, but the possible consequences flashed before me as I hurtled downwards. I was trying to access the River Garry without walking across the double track railway. I had thought that the water culverts might provide a safe passage underneath. Having located a well-constructed one about two feet wide and three feet deep I picked my way down its side towards where it disappeared in a black hole under the railway. About ten yards from this spot, a willow was half blocking my path. I was squeezing my way past when my rucksack, which was slung on only one shoulder, snagged and made me do a quick shuffle. My foot stopped on the hidden brick edge and I was spun round and propelled into the culvert on my back. Fortunately, not for the first time and probably not the last, the rucksack softened the fall. Unfortunately the steepness of the culvert and the water-covered green slime gave my feet no purchase whatsoever and I accelerated feet first towards the hole in the wall. Convinced I was going to smack my face on the wall if I sat up, I lay back and resigned myself to imminent pain. To my amazement I shot through the hole and dropped into a wet chamber, landing with my back against a wall with cold water pouring down my neck. Once again, my rucksack had been a cushion. My immediate relief at not feeling any pain was tempered by a metallic noise above and behind as my three foot crowbar belatedly followed me. Before I could move it shot over my head and landed harmlessly beside me. Phew! After a little, my eyes became accustomed to the dark. I was in a box about 8x 6x 6 feet. My route under the railway went no further as a downward pipe at the side was the route for the water. Did I have an escape route? My entrance hole was the only safe way out, but I wasn't certain I could get enough purchase to pull myself out. I wedged the crowbar across the culvert, threw the rucksack above it then jumped and pulled myself on my belly. To my relief I was able to ease out of my coffin. A few photos and I resumed my search for a safe crossing. Did I find any gold? As with so many Scottish rivers and burns, I found a few specks, but no 'wow!' pans.

For several days in Alaska's Bear Creek I had been finding some nice bits of gold, but the spot was finished and I was exploring for somewhere new.

The scene of my rapid descent underground.

The previous winter's snowfall had avalanched further upstream and had filled the valley floor with about twenty feet of hard packed snow. The creek had carved a cave through it and since it was July, the tunnel roof was well above my head. I looked up into the tunnel and thought there was potential for a new hole. Having once fallen through a drift into a stream in the Scottish hills, I hesitated and contemplated. There were other safer areas downstream and I turned and began clearing a spot 30 yards below the tunnel. Five minutes later, there was an almighty crash and the hiss of spray landing on rock and water. The roof had collapsed and lumps of icy snow up to the size of an armchair, were bumping and grinding towards me. A hundred yards downstream Colinne and the claim owner Kelly had heard the rumble. They came rushing up round the bend, relieved to see me watching icebergs drifting past.

Melting snow helped provide what was potentially my most dangerous moment while prospecting. In April 2013 I got round to visiting several burns in the Dalwhinnie area. I was desperate for a little prospecting, as distinct from gold digging, but I picked a bad day. A rise in temperature was melting the snow and the burns were rising rapidly. I nonetheless walked from the lay-by on the A9 and took the track under the railway, fording the Allt a' Chaorainn with some difficulty, the aim being to look more closely at the rock of the Allt Coire Dhomhain. There was to be no close inspection as I realised any slip might be my last. What did catch my eye in the foaming water was the number of quartz veins. I determined to come back in better conditions. In my brief absence from the car the Allt a' Chaorainn had risen considerably. As a hill walker and angler I've waded in some swift streams and rivers, but this one was now making me think twice about re-crossing. I picked the widest spot and carefully prodded my walking poles in the water. They did not go straight to the bottom but were pushed downstream. The second attempt was more forceful and I gingerly entered. My concentration was intense as a slip would have swept me into the much deeper and faster Allt Coire Dhomhain. Each footstep was carefully placed and my weight slowly transferred. It crossed my mind that it might have been safer to cross the railway upstream and risk the fine. Fortunately, the fast water had covered the area with fine gravel and there were no large stones to trip over. It was the scariest crossing I'd ever made and I was not going to cross that stream again till it was at a summer low.

Quartz veins in the banks of the Allt Coire. Seeing quartz in a new stream raises hopes. On this occasion, panning was out of the question, but even high water can be a blessing. Gold may be found in cracks above present day water levels.

If melting snow was a problem in Drumochter, new snow had posed a threat in 2011. I had decided to look again in a Perthshire burn where I suspected there might be gold. The first trip had been in torrential rain and I had come back with nothing, as I was looking primarily at the access and had gone ill-equipped. I went back better prepared. After two cold December days I knew it was there, but I had not found a really productive spot. On the third day I woke to snow blanketing the ground. Further heavy showers were adding to the cover as I wandered up the bank looking for a likely spot. Near a rapid in a rock gorge I thought I heard a crash above the sound of the tumbling water. Looking round I saw a branch twisting in the white water and assumed it had suddenly become dislodged by the current and crashed against a rock. Without further thought I continued upstream finding a small beach on a rock shelf. I immediately set up my Hungry Haggis sluice. The first panful of gravel was promising as I could see a number of flakes in the ribbed rubber at the top. Each pan-load added to my total and I realised if things kept going at the same rate, I would easily improve on my previous two days. I became totally absorbed in clearing the beach, paying little heed to the fact that my rucksack now had several inches of snow on it.

Suddenly there was a crack and a large splash seven yards upstream. Looking up I could see the large birch branch was going to drift into my sluice and tip my takings. It dawned. The earlier incident had been a branch snapping off. I recalled the winter of 2009 when the birch at my back gate was bent to ground level and I spent several days with a chainsaw clearing fallen limbs from paths near my house. I looked up. The tree three yards from me was a birch and a rotten one at that. Twenty feet above my head was a large limb with no live small branches and it was covered in snow. The sluice was quickly removed and the contents panned and made safe in the snuffer bottle. Without packing I shook the snow off the rucksack and muttering mild expletives, grabbed all my gear and headed for more open territory. My take was enough to make me think I should return and explore more. I have, but never find what I can get in many other burns. I had enjoyed the trip for a number of reasons, but it was not a place to look for gold in the snow.

Considering the number of large angular blocks I come across in possible panning areas in steeper valleys, I've never been around during a rock fall, nor have I heard any tales of accidents from falling rock. Rumbles of falling rock would make me look up very quickly. Rumbles of thunder usually give you a little more warning of trouble. When first heard, the associated lightning is well off. Nonetheless warning it is and the panner should take heed. The two pastimes which suffer the most lightning casualties are angling and golf. It's hardly surprising, as the participants of both wander around with carbon fibre or steel lightning conductors. So do panners, though I've never heard of one being struck by lightning. Only three people are killed by lightning each year in Britain, often as a result of sheltering from rain under a wet tree. Nearer one hundred die in the USA. While in Colorado looking for gold I became aware of how frequently lightning strikes take place in summer.

The first instance was at the Phoenix Mine near Idaho Springs. Colinne and I wanted to go down a hard rock mine, but on arriving we were told that the thunderstorm in the area made it too dangerous to run a tour. This puzzled me until later in the day when the tour did take place and an explanation was given.

The owners of the Phoenix Mine had followed a vein downwards, but had come to a spot where the direction to follow was far from clear. It was while deep in the mine that the owner, unaware of the lightning outside, was struck by a bolt which came through the roof and passed on downwards through his feet. While recovering in hospital he had time to reflect. Why had it jumped downward? On returning to work he concentrated his search at the point below where he had been hit. It was there that he picked up the gold bearing vein which was to save the mine. Was this chance? The guide also pointed out that the old hard rock miners used to watch for lightning strikes then carefully prospect the surface at the spot. Since metals are highly conductive, the strike often pointed out where a vein surfaced.

On another trip I had decided to climb Copper Mountain to look for mineral samples. Copper Mountain is a ski resort and along with other walkers we took one of the lifts up. The crowd dissipated and eventually we were at the top where there was little vegetation and the rocks showed greenish tinges. We found a few samples, but in my enthusiasm for looking at the rocks, I had missed the change in the weather. A rumble alerted me to a fast approaching sheet of rain and associated lightning strikes. We were exposed on a bare summit and my utterance was short, "Run!" Colinne was not a runner, but on this occasion she did. We raced down to the tree line, heading for a break which would lead to the lift. 'Keep away from the trees' I shouted in the belief that they would be

Panning need not stop because it is winter. Lack of water clarity and its force can be a hindrance in any season. On this occasion snow was breaking off rotten tree limbs.

struck before us. We reached the lift. It was stopped. Nobody was around and we were still high on the mountain. Then a door opened and an incredulous lift worker came out. He called base station and they agreed to start the lift again. We boarded and were rocketed off the mountain at over twice normal speed. The minute our feet hit the boards at the bottom, the lift stopped. It was explained that it was not uncommon to have to close the lift at short notice and that sometimes they had to evacuate walkers by bus during thunderstorms.

Steep slopes are the constant companion of the gold prospector. My pre-trip research on Alaska had put me in touch with a claim owner in the Dutch hills, just south of Mt McKinley or Denali, the highest mountain on the North American continent. The Blue Ribbon Mine was being worked in a spot called 'The Potato Patch', so called because the nuggets seemed to come in clusters and lines. This was probably due to the fact that they had not travelled far from their rock source. I was intrigued to see a working placer mine and although invited to have a look around, I couldn't dig. Not to worry, I was told I could dig elsewhere and there was also a public digging area not far away.
.
I had read warnings that the track was really rough. Even the approach from the Parker Highway at Trapper Creek to our cabin further west, was rough enough to shred a tyre without my noticing any change in the handling of the hire 4x4. Nonetheless, we set out in the pick-up and bounced our way to Petersville and beyond, spending an interesting day on the public panning area and exploring around. On return to base, I was reluctant to risk wrecking the hire vehicle on a second trip as I knew we would have to take it up a narrow steep track to the mine. Could I hire a quad bike (ATV) from the lodge? The answer was yes. Next day we set off to do the fifteen miles. The gear was lashed to the front and Colinne held me tighter than she had done for some years. The bike coped well with the rollercoaster ride and Colinne began to relax and enjoy the views of the snow-covered Mt Denali. This time we reached the base of the narrow hillside track which climbed above Willow Creek to the mine. On cresting a blind rise I was shocked to see the track a few yards ahead had been washed out and even if I braked, at the speed I was going, we were going to roll sideways into the valley to our right. I accelerated up the bank to our left, did a wall of death run above the gap and landed intact on the other side. My passenger was shaken, so much so that she refused to mount the bike on the return leg until I hit the gentler gradients of Cottonwood Creek. That visit introduced me to the industrial style mining now commonly seen on TV films from America. There was a trommel, screening plant, several excavators (some of which had seen better days) and a big settling pond for the muddy water. Later that day we panned in Cottonwood Creek finding fine gold in every pan, even on the surface. It wasn't the most relaxing spot as there was plenty of signs of bear activity so we kept the quad close and well warmed up.

I had read up on bears before going to Alaska, but had no weapons with me other than advice. Some carry firearms, as I mentioned earlier, but bear pepper spray is the recommended weapon for those going into bear territory. It is proven to be twice as effective at deterring a bear if used properly. My tactics were to make plenty of noise so that any hidden bear was not surprised. We also avoided areas where there was limited visibility. When I asked Kelly what he did when he encountered bears in Bear Creek he said "I speak to them". It's a technique I have seen used in a number of films. "Hey bear" is the usual opening gambit. Kelly held his hand up in salute. I didn't get a chance to try this out, but would now add a capsaicin bear spray to my kit if I went back. I would also make sure I knew how to use it quickly.

Fortunately bears are not a problem in the woods where I now do most of my panning, but there is one creature which can drive you off if not well-prepared. I refer to Culicoides impunctatus, the midge. Almost all Scottish streams are plagued by the little horrors on calm days from spring till autumn. Although there are various deterrent sprays, I prefer to have a green midge net, long sleeves and gloves on. Leon says the best way to find gold untroubled by midges is to snipe. As long as you don't suck them down your snorkel, there is nothing for them to latch on to.

While panning in Colorado, I bought a book called *How to Shit in the Woods – An environmentally sound*

Finding gold in the 'Allt G' in what Jamie Shepherd described as "midge hell".

approach to a lost art by Kathleen Meyer. It is full of humorous incidents, but has a serious side. It warns of the spread of Giardia, a parasite that will cause you internal problems. Since then I have stopped drinking even the clearest looking Highland stream other than through the survival straw. It has a proven filter and being small and light remains in my rucksack survival pocket.

Panning can be a thirsty business. Carrying liquid can add to your weight. It is tempting to use what may seem clean stream water. The following tale will make the reader think twice. When I met up with John Wilkinson in May 2014, he had just come out of hospital. The news that he had resorted to the NHS surprised me as John had seriously gashed himself at Wanlockhead late one night and rather than heading off to A and E in Dumfries he had sewed himself together in his caravan. What was so serious that he had to be admitted to hospital? They had discovered he had a large liver fluke and had started his cure just in time. John had always quenched his thirst straight from the streams in which he panned.

When sniping, it is possible to belt yourself in the face with a tool since you are often working very close to the rock. In the early days of his sniping career, Emil Heinneman from the Netherlands did not use a facemask. In order to keep both hands free he held a 'beanseen' viewer in his teeth. What he had not allowed for was the pressure of the current. The result was a need to visit a dentist for major repairs. Having said all this, gold panning is not a particularly dangerous pursuit, but it is often practised in more remote areas with no mobile phone signal. It pays to go with a companion and if you don't, you should take some emergency gear with you. The survival pocket of my rucksack contains more than the survival straw. I carry a silver thermal blanket; painkillers; plasters; high energy food; a torch (I usually have an underwater one with me anyway) and of course my mobile phone, in the hope I might get a signal.

As I get less nimble (a euphemism for being downright stiff at times) I take walking poles which are an asset when crossing larger streams, travelling over rough ground, prodding for footholds on steep slopes and taking weight off my knees going downhill. If the zip pull on my drysuit is sticking out, Leon will remind me of the potential for snagging, but as yet I have not made a cover for my T handled pump which, when attached to the side of my rucksack, frequently snags on tree branches. It has dumped me on the ground more than once. I'm also far more aware of green and wet rocks these days, having prostrated myself on several occasions without so far breaking anything. Am I tempting fate?

I notice that the miners in the programmes filmed in North America now all have hard hats and hi-vis jackets. Gold panning has its dangers, but I trust we are a long way from needing that.

Emil Heinneman, Mike Fisher, Leon Kirk and John Wilkinson beside the Mennock. Whether you work alone, in a team, or as individuals in a group, the object is to enjoy the day finding gold.

GOOD DAYS and BAD DAYS

'The fun really is in the process of finding the gold, not in the gold itself.'
Charlie Smart

The angler and the gold panner have a lot in common. They set out to capture an item and hopefully at the end of the day have enough to proudly display the successful outcome. The angler is very much at mercy of the weather as well as the mood of the fish. The panner can be caught out by the weather, but inanimate gold does not change its position or difficulty of capture on a daily basis. He or she has fewer excuses for a poor take, especially as since the last trip, there has been time to work out exactly where the gold should be and how to set about collecting it.

I have already given instances of weather causing me problems in my quest. Another which springs to mind was when I drove the hour and a half to the Garry, confident that because my local burns were low, I was going to have an enjoyable day. On arrival, I found the watercourses so high I couldn't even cross a normally tiny tributary to get to the spot on the opposite bank of the main river from where I had been working. The river itself was reminiscent of the rapids on the Colorado. High water can often make you re-evaluate a river and look elsewhere to your advantage. On this occasion I had no alternative but to turn round and go home.

An instance of knowing where the gold was but landing up with a poor take happened to me on the Kildonan. I was looking forward to finishing a spot I had been getting ever-improving returns from, but had had to abandon. On returning I found it had been thoroughly cleaned out in my brief absence.

If you go to a new stream and have had no time to think about where you will dig, then luck can be involved. A vivid memory is of spending a day with a group on the Shortcleuch when everyone else found at least one nugget and I eventually got one very small flake.

As a salmon angler I carried at least three reels, each with a different type of line. My fly box also contained anything from a five inch Collie Dog to a very small Silver Stoat. This allowed me to try to tempt fish in a great variety of water levels and temperatures as well as pandering to the fish's whims. Normally I go gold seeking similarly well-equipped, having thought out beforehand the best method for the water involved. Imagine if you were told that you were joining a party to fish for trout only to find they had changed their minds and were going salmon fishing. It's not impossible to catch salmon on a trout rod and with trout flies, but it makes it more difficult.

The gold panning equivalent happened to me at the 2015 BGA weekend in Tyndrum. A certain friend had said he wanted help to photograph some new sluices. I packed cameras and waders, not a drysuit and sniping gear. Only on arrival did I find gold fever had returned and the rest of the crew were going sniping. They all did well in deep water, while I, without the right gear, did not. I later returned to the river and rectified my error. I also vowed to pack my vehicle with the full range of equipment whenever possible.

It is one thing to have a poor day at home, but to travel across the Atlantic anticipating good pans of chunky gold only to find the area off-limits is a bad start to a panning trip. We were on our second day of our third trip to Breckenridge, Colorado, settling in to removing gold-filled clay from the dredge tailings of French Gulch. Through the cover of willow trees a voice from the dirt track began shouting "Come out of there, I know you are in there, I'm going to call the sheriff". On emerging, a guy on a trail bike asked us to leave, explaining he was the caretaker. The old gold dredge tailings had been bought for 'real estate' (housing to you and me) and the new owner didn't want it known there was gold. I explained how far we had come and that in previous years umpteen panners from all over the States had worked this spot. He was sympathetic and relented, but rather than get him into trouble, we left, after an amicable chat. We did not find as good a spot again. On the last check on the internet fifteen years later the housing was nowhere near the hotspot. I could go on, but I go gold panning for enjoyment and try to forget bad days or setbacks. It is far better to remember the good days. What makes a good day out?

Andrew Winter is a very successful gold seeker who enjoys both amassing gold, but also prospecting new streams. His account of finding the nugget he called the 'Tyndrum Tiddler' is full of the heart-racing excitement that a good nugget brings:

"It was in July 2010 when a group of us decided to head to Tyndrum for a three day weekend at our usual spot on the Crom Allt. This was a spot we had been working for some time during 2010. We weren't the only ones working this spot and it appeared as though we had been swapping weekends with someone else who we never managed to meet. Still, there was plenty to work and these prospectors seemed to leave the best spots alone.

Day one saw Robbie Falconer (Robthor); Frogesque; Pedrog; Corbie and myself working in different parts of the area. Corbie and Pedrog continued their work on their previous glory hole while Robthor decided to work on the upper section of the area and dig deep for nuggets. Meanwhile Frogesque, who was working a new spot opposite, was also intermittently testing some of the tailings that Robthor was vigorously discarding.

After a good day's work, we were set up for and looking forward to the following day. After the usual post panning fare of food, beer and banter, we headed to bed to recuperate. Unfortunately no one had told the dark overhead clouds of our plan. We awoke in the comfort of our hotel to find there had been substantial rain overnight and on arriving at the river found it in full spate and far too dangerous to return to our spot. Not to be put off, we headed upstream to a more accessible area, one known to have produced nuggets in the past.

It was a wet day to say the least and we found little gold. Not only that, my bucket had rusted through and developed several holes and I had managed to throw away my riddle along with the stones I was discarding. On the positive side, the water did drop sufficiently to give us confidence that by tomorrow we would be back in our usual spot. It also crossed my mind that it must be my turn to find something big. I was banking on things happening in threes.

That evening was a repeat of the previous one but there was no overnight rain. In fact by morning the sun was shining and the Crom Allt had dropped to a low level. Unfortunately, the spate had been quite substantial and had filled every spot we had been working. It had made a fresh-looking area and we were guessing exactly where we had been working. There was no sign of my sacrificial riddle, but I had replaced my rusty old bucket with a nice new shiny one. We all decided to start working the spots which we thought we had left alone. Robthor and I agreed we would work two areas next to one another with the intention of joining them. This was an area of hardpack which I had tried before, finding it difficult to remove. It wasn't long before we were all finding gold and the area started to resemble the workings of panners 'going for it'. Robthor was getting down to bedrock and in my attempt to keep up I had to resort to getting my facemask on, plunging my face in the water and with the greatly-improved visibility, filling the bucket with the larger stones. It was during one of these clearing sessions that out of the corner of my eye I caught a glimmer of what looked like yellow shiny mica. This was no small piece of shiny stuff but as is often the case with glitter or yellow spotted under water, it turns out to look very different on the surface. I picked up the shiny yellow object, placed it in my glove and drew it closer to my mask for inspection. I was still unsure of what I had found until I took my head out of the water, removed my mask. There was no doubting what was in my hand. At this point I started shouting 'nugget, nugget'. The rest ran over to see what all the commotion was about. I carefully handed it to Pedrog who then handed it round. With this, I was so worried about it being dropped that I could not help myself from clasping my hands together under each hand holding my prize. I will never forget the look on Corbie's face when he asked Pedrog if he had found it, to be told "No" in a disappointed voice – they were working the same hole and sharing everything they found. My hands shook and I giggled with joy and nervousness coming out with the classic truism, "it's too big to snuff up into my snuffer bottle". While repeatedly uttering these words, I unscrewed the lid and popped it in the bottle. I wasn't the only excited panner muttering a statement over and over, but in the case of Frogesque it was "You jammy git". By this time Robthor was wondering how many nuggets he had thrown away during the morning clearance of the surface gravel which he had been throwing downstream towards my excavation. The Tyndrum Tiddler weighed 5.41 grams and made up my total for the weekend to nearly 8 grams, my best to date."

There are a lot of lessons to be learned from Andrew's story. On reflection, the nugget had come from surface gravel about one foot down, upstream from a large boulder which separated Robbie and him. Robbie will never know if, or how many Tyndrum Tiddlers he threw away that day. Nonetheless it makes for a great conversation topic when we get together. It also makes you wonder about how much gravel you should throw away without testing in the quest to get to bedrock.

Fact: Nuggets are given names - the Welcome Stranger (largest ever found), the Hand of Faith (largest in existence), the Honourable Roddy (New Zealand's largest). What's yours called?

In June 2015 I had an evening call from Leon. He was far more excited than usual, but at the same time I sensed he was in shock. He just had to tell me what had happened on his last instruction weekend. It wasn't long before it was all across the media and this is how Leon's story of the best find of the last half-decade appeared in the BGA newsletter after my editing:

Back in January I was contacted by John, a Canadian, who wanted me to spend three days showing him panning in the area around Leadhills and Wanlockhead. He had done some panning in Canada, but had very little experience. Occasionally he had taken a three hour drive from Vancouver to get away from the pressures of his business. He enjoyed the fresh air, the chance to see wildlife and trying to pan for a few small flakes of gold.

His first day was a nice bright dry day, so we decided, after a coffee and scone at the Museum of Lead Mining, to head to the top of the Mennock Water to a spot where I had pulled out a 0.4gram picker just before the 2015 BGA championships. We had a good day just chatting and enjoying the tranquillity of the river and managed to find a few flakes of gold, but nothing to write home about. On day two we decided to continue our morning ritual of coffee and scone and this time go to the Wanlock Water. After parking the car we walked downstream for about a mile, finally stopping at what looked like a promising spot. This was to be one of the hottest days of the year and the sun made it uncomfortable prospecting. We ended with barely enough gold to squeeze into a midge's bum, not that we tried. For his last day, we had arranged to meet up with Tony and Pat, friends of mine who were in the area. They had come up from Lytham St Anne's and were staying in their motorhome, having come north for a few days panning. They had not had much luck on the previous few days. I decided to go to a spot at the top of the stream where I had found some nice pieces of gold a few weeks before when filming for the BBC One Show. After doing a little sampling, I gave a quick demonstration of how to snipe gold from the bedrock with the help of an underwater viewer. Impressed with this new technique John decided to have a go and went off with a viewer and minimal tools. In a matter of minutes he came back with something in his hand asking cautiously, "Leon is this gold"? His body language suggested he was prepared for the disappointment of a negative response and was prepared to toss it away. At a distance as he approached me, it looked like a yellow stone and I almost said "Chuck it back you silly Canadian". As he came closer, the colour was still not jumping out at me, but I could clearly see the surface had a metallic look. When he put it on my hand it was still not clear that it was gold, as my 5mm thick gloves masked its weight. I removed my glove, dropped it in my hand and my heart nearly stopped. Oh yes, it was gold alright. It was the biggest piece of gold that I had ever seen come from any Scottish river and I have been doing this for nearly 20 years. Tony and Pat were standing next to me while all of this was going on. I will never forget Pat saying quite seriously to Tony as though this was an everyday occurrence, "You never find pieces of gold like that for me". The nugget was slightly duller than the gold I am accustomed to in the Mennock, but the dull algae look soon wore off with handling it. John showed me where he had found it. The nugget had not been in hardpack. It had been lying about a foot below the water's surface in a hematite-filled bedrock trap and covered with a stone.

Two things occurred to me that day. The first was that the nugget would not have gone through a standard gold panner's classifier. It could have been dug in the past and thrown back if the panner had not checked carefully. The second was that this nugget came with no warning, just like my Pikoe and Pedro nuggets which are both over half an ounce.

After a little discussion I decided that our hobby had been plagued with bad publicity over the last few years so we would let this feel-good story go to the media. We headed for the Museum of Lead Mining to break the news. A small crowd gathered round as it was carefully weighed – a whopping 18.1 grams. Then we contacted the media, stipulating two conditions. John's name and my business were not to be mentioned in the story. Word spreads fast and before I left the museum, I was called to the phone to do a live radio broadcast for BBC Radio Scotland. As you can imagine, the beers were flowing that night at the Abington Hotel where John was staying and some of the staff were lucky enough to get the chance to hold it before its long journey to Canada.

Well, as they say, the rest is history. The story went viral and within a couple of days would be printed in most UK newspapers and many others around the world. The write-up was all positive and gave a new shine to our hobby, one result being the museum was busier than ever.

With every action there is always a reaction and in this case, like Othello, some had fallen under the sway of the green-eyed monster. This was not from the media or the general public but from a few envious panners trying to disparage the nugget and the circumstances of its find. After a little thought and some telephone calls, we decided to call it the 'Envy' nugget.

John works in the movie business, hence the reason for his wish to be kept out of the story. The benefit of this is that as he specialises in 3D printing, he has made a copy of the 'Envy' nugget that is now on display in Wanlockhead's museum for all to see. His abiding memory of the whole experience was the astonished look on my face when I realised what he had found. For my part, I consider him the luckiest panner I'm ever likely to meet.

This seems to confirm Andrew Winter's feeling that nuggets can be carelessly thrown away when panners are discarding gravel not expecting anything of that size in that spot. Some of Bob Sutherland's larger pieces were known to have stuck in the classifier. Although many good pieces of gold work their way into the bedrock, Leon's 16.5 gram Pikoe nugget was found in gravel which could have been moved in the next big flood.

A replica of the nugget was sent to the Lead Mining Museum and put beside John Hooper's find. His beautiful gold in quartz nugget immediately caught my eye when I first viewed the display. When I first met John in 2011, he was the ebullient and energetic president of the BGA and the driving force behind the annual staging of the gold panning championships. He had started panning in the Leadhills area in the mid 90s. About two years after starting, he was scraping away gravel on the Mennock one day, when he spotted a pothole with the aid of his viewer. In the hole he could make out roof lead. He pumped out the contents into a bucket

John Hooper's 6.1 gram Mennock nugget on the left next to the 18.1 gram Mennock 'Envy' nugget. Photograph taken at the Museum of Lead Mining, Wanlockhead.

with a sieve, shaking it to get the dirt to drop through. He was about to throw the sieve contents away when he noticed the quartz with gold stuck in the mesh. He did pan out the bucket, but couldn't wait to show his find to someone, so he headed down the glen to Charlie Smart's at Mennock Cottage. Charlie got very excited. "You know what you've got?" said Charlie. The 6.1 gram nugget was the biggest from the burn in the last 50 years.

My own nugget story (its size is relatively insignificant compared with those above) was played out on the Crom Allt in 2011. I had met up with John Greenwood and we had decided to work close to one another, but by different means. John had set up a sluice, but I was panning. He immediately got one or two good bits showing in his sluice. We were well up the burn and I was against a rock wall near where I had found good bits before. On using my long-handled shovel to take out gravel, the tip kept hitting a stone. I forced the blade between the offending rock and the vertical rock wall and levered. It took some time before I felt movement and the rock loosened up. I tried lifting it and realised it was not very big. Two hands pulled it free and I was about to toss it away when I noticed it had clay on it and it was not rounded. I put it in my pan and gave it a thorough wash, immediately swirling the grey water round. There was no gravel, but a dark stone was stuck there resisting the water movement. Against the light it looked like a lozenge. With fresh water I washed it again. The water stayed clear, the stone stayed put. It was at that point I looked more closely and realised I had a nugget, with rusty iron residue on one side. The rock I had removed was in fact bedrock and the nugget had been in the cleft where I had stuck the shovel point. I called John up and he got very excited. Where there was one there could be more. Before I had a chance to look, John had asked if he could use the shuftyscope to look at the spot. He knelt down and peered in shouting "I can see another, can I pick it out. I think there might be two". I let him enjoy the moment and he managed to pick out another much smaller piece. The third piece came out with the pump. In the excitement of the moment, I didn't see either bit in situ, but in a matter of minutes I had over four grams of gold. I had nearly thrown away a three and a half gram nugget, but for a little voice saying to me "Always wash any clay off rocks".

Mike Fisher is a quiet gold-seeker who had suffered a lot of disappointments in his early career. On one occasion I remember Terry Mahoney inviting him and me to leave the gold panning championships and go to the Glengaber for a day. I declined as I was down to compete and to help, but Mike accepted the invitation. That evening, Mike was a bit glum. He had sniped across a crack, getting nothing and had given up. Terry went in and removed one or two little pebbles at the end of the crack and took out a handsome piece. This did not help Mike's frame of mind and the feeling that he was never going to have any success. Nonetheless, he drove up from England in May 2011 to join a group of us who were going to try the Shortcleuch. Leon was aware that Mike had not had any great success and on reaching the burn's edge pointed to a rock area which he knew had a crack across it. The rest of us headed upstream. We all gathered at lunchtime except Mike. None of us had a speck. We were eating when Mike appeared and dropped a good flake into a pan. We were impressed, but he then rattled the bulb of his snifter and poured out a three quarter gram picker of beautiful colour and form. No-one did better that day.

I once spent half an hour trying to remove a nugget from the bottom of an Alaskan stream, only to find it was a lump of pyrite. Had I been sniping with goggles, it would have been obvious at the outset. Colinne had a similar experience on the Crom Allt, similar, but not quite. We were working below the graveyard on a sunny day near a spot that had been good to us the previous year, 2003. As so often happens, floods had changed the distribution of gravel and some bedrock had become exposed. In the bright light Colinne spotted what she thought was a nugget glinting on the bottom. Fearing ridicule if she was wrong, she said nothing and tried to pick it up. It would not budge. She went for the viewer and had a closer look, still saying nothing about her mounting excitement. Only when she was convinced that it was gold and she might need some help to extract it, did she say, "Alan, I think you had better come and see this". With chest waders, I was able to kneel and get closer. What I saw was a thin

Two half gram pickers plus over a gram in the vial, all from the same crack. On this particular river, I could not find another gold filled spot upstream or down for over one hundred yards - some rivers can be patchy.

vein of white quartz about a foot long, running across the grey schist, with worn gold pieces sticking out at intervals. I tried to remove it as one piece, but the rock was too friable and broke. Nonetheless, we retrieved every last piece. What Colinne remembers most, was the fact that I said "Colinne, you're brilliant." She maintains to this day she would rather have picked up a nugget, but the pieces are almost unique and have been much admired. I have since estimated there is over a gram in the bit showing most gold but there are about twenty other bits.

Bob Sutherland's glittering career reached its zenith on the West Water where he discovered that the pieces were so large that the odd one could stick in the classifier. On his first day, large pieces were coming out thick and fast and instead of putting them in a receptacle, they were laid on rocks around the working area. Bob was so engrossed in pumping more that he left the rounding up of his finds until the light had gone. When he came back next day, he discovered there were still lumps of gold lying on some of the rocks.

Leon Kirk has not infrequently complained in my presence at the start of a day, "It's a while since I've had a five gram day". I assumed from this that five grams was what he considered a good day. Having seen his nugget collection on a number of occasions, it is hardly surprising that five grams would be a good benchmark for Leon. Five grams has always been beyond my expectations for a day and if truth be told, I've never achieved it. On our first trip together to the Suisgill, our first day was poor as neither of us hit a good spot. I had been getting a steady trickle of small flakes, but when I reached bedrock, it was smooth and hard and gave no chance for gold being trapped. I had only half a gram. Day two gave slightly more, but Leon was not happy with his return. On the third day we were well apart and he still didn't look happy when we met up. We motored into Helmsdale to eat that evening and settled on having fish in La Mirage. It was then that he produced a glass vial and rattled it in front of us. Both he and Jamie, who was in on the deception, were beaming. He had his five gram day, including a one gram nugget.

Jamie himself far exceeded the five gram day at one stroke in 2013. He recounted the story at the BGA meeting at the Wigwams in Tyndrum in October 2014. James Linnette (Helicopter James) had been developing a hole in the gravels of the Crom Allt and Jamie had agreed to join him, but had got delayed helping Lobster James further upstream.

Helicopter James had reached a stage where further gravel extraction was impossible without moving a large boulder. Try as he might he could not budge it and was glad to see Jamie arriving. Between them they moved the boulder out of the way. Jamie, rather than using a viewer, has goggles or a facemask almost permanently slung on his head when panning. With the boulder out of the way, down came the mask and within seconds, Jamie was gazing at the exposed gravel. James Linnette was getting ready to snipe when Jamie popped up saying, "You're never going to believe this". The object of his euphoria was a thirteen gram nugget. He showed it and described the find to all assembled at the BGA's September weekend in Tyndrum in 2014. Give James his due, he did not let this affect his relationship with Jamie.

One panning experience I remember vividly was a bright day in Colorado. Instead of returning to the hole we were digging in the dredge tailings of French Gulch, I took a tour up the Swan River first. We arrived at our digging spot an hour later than usual to find another vehicle there. The three were unloading formidable equipment, including a power generator and an industrial-sized vacuum cleaner. We headed for our hole on a ridge beyond the willows, but on arriving found pans. I had left a newly-carved stick in the hole the previous day as an indicator it was still being worked. The three appeared carting their gear and warned us off. I politely pointed out that we had developed the hole and were not finished. It was made plain that since they had arrived first that day they had claim to the hole and if necessary they would use their superior weight and numbers to enforce it. We moved to another ridge about 40 yards away and began removing the top stones to get at the clay which became more concentrated about a foot down. All the time we kept an eye on the three. With the generator running they vacuumed the hole, eventually tipping the contents into a pan. There were nods of approval. We noticed they were not clearing away the stones, but relying on the power of the vacuum. In the meantime we had filled and reduced a couple of pans and had realised they had done us a service. We worked on, giving no sign of how well we were doing. We left our departure till well after the 'claim jumpers' had gone and were more than happy with our take from the new hole. Next day we made sure we were there early, but we did not see them again.

WHAT MAKES A GOOD DAY'S GOLD HUNTING?

In 2012 I asked myself the question 'What makes a good day out?' I answered it for an article in the BGA newsletter as follows:

If you travel some distance to the Gold Panning Championships and have the time, it's as well to spend a few days exploring the streams in the Queensberry Estate and Hopetoun Estate permit areas. After the 2012 event I had a day on the Wanlock Water where success in an otherwise fruitless day was measured in terms of sniping a bigger flake than I'd hitherto taken from that stream. I had to be home next day but thought I could manage a couple of hours before driving north. Anticipating an early start I had said my farewells the previous evening after singing a few songs in Mike and Colleen Jones' hospitality tent, i.e. the extension to their caravan. With a last wave to Karl-Heinz Heidler, I headed for Leadhills next morning and took the Elvanfoot road to the Longcleuch.

A couple of years ago I had walked up to the source where Beavis Bulmer had mined for gold in the 16th century. I had sampled the gravel there and had to dig a hole and let it slowly fill with water to enable me to pan out a few specks. On my way back I saw a likely spot and immediately got a small flake. This was followed by disappointing pans which nonetheless were filled with haematite pellets. I didn't manage to reach bedrock before I had to leave and was convinced I would find more gold if

I got deeper. This time I would snipe on bedrock, despite having been told it was not a good sniping burn. I was soon to find the rock was hard, the cracks few and in many places the water was so low bedrock was three inches from my nose. Fortunately I had attached a pan and pump to my pack, but after an hour I had four tiny specks for my efforts and I still couldn't find my haematite spot. I headed down again and looking from above I thought I recognised the spot, but there had been a change. The boulder was still there, but most of the gravel was gone and the bank was eroded. I had twenty minutes left of my planned two hours. The first pan produced a load of haematite, but no gold. The rock was smooth midstream but where the bank had been eroded there were cracks. I eased away the loose rock but there was no depth of water to use a pump. I considered making a dam, but this was out of the question. Then I remembered I had thrown in a trowel, not a tool I normally carry when sniping. I was able to scrape up small amounts of a milky clay and gravel.

The first cupful produced a bigger flake than I had found in the Longcleuch to date. Two of the next three gave even bigger bits of very bright gold. I was to regret not carrying my pump-reducer-nozzle which can lift dirt in less than an inch of water and often produces a good clean-up. An hour had gone, it was time to go, I hadn't a gram, I hadn't a nugget and I hadn't even a big flake. In fact the take wasn't worth weighing. So why was I on such a high? I had taken more than on the previous trip, the sun was out, the primroses were still in bloom, little orange butterflies were everywhere (Small Copper?) and curlews were making their iconic call. For me there is more to gold panning than finding large amounts of gold. Albert Morris the former *Scotman* newspaper journalist quotes G.F.S. Adamson (Fred) who wrote *At the End of the Rainbow*, "Fred told me it didn't matter if we didn't find gold. It was all a great way to spend a day doing nothing". He had spent many such days. Albert finished the article about his gold-less day panning with Fred as follows, "The day had cleared up like my cold; the sun shone over the rain-dropped-jewelled fields, and the river sparkled brighter than the counter display of a diamond dealer. Somewhere, the liquid song of a damp lark was heard. That was the real gold for me".

Albert and I had something in common, but I did need to find gold as well. Not everyone will consider the above a good day, but I recommend you look at what grows and lives around the stream bank as well as studiously observing and processing the gravel and rock. Nonetheless finding gold is the primary object and what fires the gold panner. It is often hard to find a really good spot and it is getting harder by the year. To give one away totally inadvertently can stick in the mind for a long time. One instance comes to mind. Leon, George Marshall and I had been digging and sniping in the Suisgill finding some bits, but nothing to get really excited about. We were walking back to the vehicles with me at the tail when I diverted to look into a pool. On catching up I informed the others that I could see good cracks and I thought it might have some potential. Next day Leon and I left George at the cottage and headed back to where we had worked the day before. George did not appear. We had an average day and on dropping down past the pool I had spotted, here was George packing up and being straight-faced and non-committal about his day. It was decided to compare results and Leon showed one or two good pieces in his pan. When George emptied his snuffer and swirled away the heavies our faces must have been a picture. The top edge was littered with large flakes and small nuggets. "Eat your heart out boys" said George.

The point where the Longcleuch goes under the old railway. Beyond is the valley of the Shortcleuch.

The Longcleuch, a tiny stream with a little gold.

Blink and you could miss it. The haematite is obvious, but there is a small flake. It's where you might expect it near the edge. I had little gold but an enjoyable day for many reasons.

George Marshall shows who's had the better day. Look at the pan, not the size of the smile!

Next day we were all in the pool and found a few more good pieces, but not the haul that George had taken. A year or so later I was sitting on the bank opposite the main crack George had worked. I was out for a walk and had no gear within quarter of a mile. As I gazed into the water thinking of how I'd put George on to a good spot, I spotted a glint near the bank but in deep water. Shining items under water often disappoint, but I could not ignore it so I went for a pump, a pan and waders. One draw and the item was dumped in the pan. For once, I was not disappointed and I had a good picker. How had we missed it, or had it been dumped there in the intervening time? It was more likely the latter.

More recently it was George who gave away a good spot inadvertently. The autumn BGA meeting in 2015 had a group of us, including George prospecting in Glen Lochy. When I arrived, Mike Elliston was highly visible in his yellow and blue dry suit digging in the middle of the river. I had not taken my suit and was resigned to shallow water crevicing in waders. I had offered to take Tim and Myles Light along and we moved upstream so as not to crowd those already digging. Unfortunately, although the rock area I had chosen looked good, it was disappointingly devoid of gold. In the meantime Mike was finding little. George had said to him "Go and dig beside that blue rock." Mike duly did and after a while was beginning to see signs in his sluice. George did not do badly for his advice. At one point I heard a shout and looked up to see George motioning to come and look at what he had uncovered in a deep V-shaped cleft. James Linnett and I gazed in while George, with difficulty, cleared away gravel which kept drifting over the spot he had already cleared. Eventually he revealed two good pieces, the larger over one gram.

My second spot turned to be marginally better than the first with one picker and a couple of flakes for a couple of hours work. As the light went, we drifted back to where Mike was still filling his trug with gravel. We all gathered round his sluice while he started to trowel the contents into it. The dirty water which had been coming out of his pump was the obvious result of being into good gold-bearing hardpack. My face lit up and I congratulated him on showing the best display of gold I had seen in a sluice feed tray. He looked up grinning. Everbody got excited, but also helped tidy up his gear as the light died. James Linnett was entrusted with panning the sluicings while Mike grabbed his scattered gear. One piece was around a gram in weight and guesses were made as to whether it was above or below.

That evening Mike and family appeared at the Wigwam hall before Dr. Neil Clark's talk. I called on him to show the assembled crowd his day's take. A proud Mike produced a vial with over four grams, including the piece which had indeed exceeded one gram. Everyone admired and congratulated him, but I was not expecting what happened next. An excited Mike said to me that what had made his day was the look of joy on my face and my congratulations on his success as he fed the sluice. What had started this euphoria? It was George pointing Mike to a spot he thought might give him a little gold.

Better was to come. I realised that Mike was in for a second good day on the Sunday. I had to head home having a gold-less prospect in Glencoe in the passing, but I did hear on the grapevine that Mike had over nine grams. I did not hear the other bit of good news from the Sunday, until Tim Light wrote his account of the weekend in the BGA winter newsletter 2015. Much to my disappointment, I had not helped to give Tim his first good gold day. That was to come next day. He, son Myles and James Linnett teamed up to work a hole close to Mike which had been vacated with the departing digger giving his blessing to its further exploitation. James went to help Mike, Tim pumped and Myles fed a large sluice which had been carted in by George. Tim and Miles could see they were on to gold, but the hole was so deep that Tim was shipping water down the front of his waders, a feeling he did not enjoy. With the weather cooling, the clean-up began. They were enjoying the experience of seeing 'ten to twenty specs appearing on the top line'. Tim was picking out the stones from the remainder of the pan when he noticed one of the black stones was much heavier than he would have expected. He asked George what he thought it might be. George's experience came to the fore and he rubbed the 'stone' on the top edge of the pan. His face broke into the Marshall grin. What Tim had just about thrown away was a dirty 2.35 gram nugget.

What was a very productive spot on the Suisgill. Gold often settles in greater concentrations in the gravels on the downstream side of large boulders.

In all they had 3.6 grams for their day. I have since worked in mid-stream near this spot taking a few small pickers which could only be initially identified by their slow movement in the pan, because they were coated in a black deposit.

Mike's spot was close to a tree growing on the bank. Tree roots close to a river frequently put out roots into the rock or gravel in the stream. They open the rock, making ideal traps for gold or help to bind the gravel and slow the current, again creating good gold-depositing conditions.

Colinne was working a similar spot on the Crom Allt where tree roots had grown well out into the gravel of the channel. On this occasion she was digging by herself, although I was not far away instructing a Lomond and Trossachs National Park ranger. From time to time she would look to see if he was watching and if not, would quickly turn her pan in my direction to show she was getting good returns. I was helping the ranger to find his first gold but not at anything like the rate next door. At the end of the day she had over two and a half grams of fine gold.

The gold totals mentioned vary considerably and must be considered in the context of the places being panned, the age and experience of the panner and the conditions at the time. As an angler I always thought that a 1lb trout from a burn that averaged ½ lb. fish was a fantastic catch just as a 2lb. Spey grilse was almost as great a disappointment as a blank day. The same is true of gold panning. Despite the fact that I was panning for a few days a year when Bob Sutherland was developing his expertise every weekend, I never had the sort of days that Bob was having. On visiting the Scottish Gold exhibition put together by Dr. Neil Clark at the Hunterian Museum in Glasgow in 2014, I took a note of some of Bob's finds on display.

Saturday 18.9.93 – 11.26g
Sunday 12.9.93 – 19g including 4 nuggets of 2.73, 2.03, 1.83 and 1.03g.
Sunday 30.10.93 – 6.9g nugget
I also noted four filled 1 oz. vials.

Bob told me that over six years he had taken over 2 kilos of gold including what he thought was the largest nugget taken in Scotland in the last century. It was 21.35g. Terry Mahoney had one of 21.98g, but Terry, being Terry, didn't like to disabuse Bob and steal his glory. Either way both prospectors had some excellent days.

James Linnett has been an enthusiastic and fast learning newcomer to the hobby. I remember meeting him a week or so after he had found his first 1 gram nugget. He was still on a high. It had been taken from a crack close to where Leon had been working and when he had shown it to Leon, the latter's expression had obviously made an impression on James. "You should have seen his face" he kept saying to me. Other people's good fortune in the gold burns can often leave you looking envious or astonished.

There is no-one more enthusiastic about finding gold than Vince Thurkettle. He will happily talk about his experiences and give the benefit of his experience to any interested person. In one chat when I had no equipment to record the details, he told me of two days which he remembers well. The first was in Wales where he had gone into an abandoned spot and taken over 5 grams. The second was in Aberfeldy when it was open to panning. He had looked under a boulder where there were signs that it had at some time in the past been worked. He dipped his face in the water twice and got that tingle of excitement. The gloves went on and he told himself "Don't get excited". What he pulled out was a 4.8 gram nugget. What did stick in my memory was what he was leading up to. "When you think you've finished, begin" he said. Good advice for any gold panner. Too many abandon a spot before they have worked it to completion. I know, I have been guilty of it myself.

These were excellent days for Vince, but I know he still gets a thrill from finding evidence of gold, however small, in a new stream. Whatever excites you about this hobby, be it a large total take, a new find, a day in good company or being out in an environment rich in nature, you are not going to get all of them every day. In fact you may be unlucky and not get any of them, but the thought that things will improve next trip is what drives gold seekers on.

Nothing large, but a good return from one pan. Don't be tempted to waste time at the river separating the gold from the other minerals. Collect both in a snuffer bottle and clean it later.

Results from a good trip despite water temperatures just above freezing. This was sorted each evening by a roaring fire. Some prefer to dry their gold. Clear vinegar heightens the lustre and kills algal growth which develops in water.

READ ALL ABOUT IT

'Once a newspaper touches a story, the facts are lost, even to the protagonist.'
Norman Mailer (1923–2007)

Gold stories not infrequently reach the press. The Kildonan Gold Rush of 1869 was well reported. R.M. Callender and P.F. Reeson researched the archives thoroughly and their book, *The Scottish Gold Rush of 1869*, has numerous press quotes from that time. After the Duke of Sutherland had closed the diggings, the *Inverness Courier* continued to report the finds of prospectors who had moved into areas outside his control. These reports are available from Inverness Reference Library and I have used paper copies to pinpoint streams to try.

Even today British gold stories regularly make the papers and over the years I have built up a collection which still make interesting reading. The oldest of these is from the *Scots Magazine* of September 1961 entitled 'The Gold of Kildonan' by Lachlan Ross. It is a potted account of the gold rush. Bearing in mind the date of this article, I now wonder if it was on reading it, that the idea of looking for Scottish gold occurred to me.

A newspaper article from *The Sunday Times* of 11th August 1996 certainly caught my attention. It was titled 'Huge gold deposits found in Britain' and was written after the British Geological Survey's four year Midas Project was released to those beyond the mining industry. The first paragraph certainly grabbed the attention. "They are calling them the new Klondikes. Geologists have found huge deposits of gold that could be worth billions of pounds underneath cities, towns and tracts of countryside across Britain". A quote from the then head of the BGS mineral research group pointed out that some areas were so rich in gold that it had seeped into streams in sufficient quantities to be panned. There was a picture of a young lady (described as "Klondiker") panning a stream in the Forest of Dean.

The other illustration was a small-scale map of Britain showing all the gold localities. They extended from Cornwall and Devon in the south to Kildonan and Gairloch in the north. Twelve red hammer symbols marked 'goldmines'. Most of these were not existing mines, but potential goldmines. Over a quarter of a century later none of them, existing or potential, was producing. The article finished with reservations about mining from conservationists and the Scottish Landowners Federation. At the time I plotted the Scottish references on larger-scale maps to extend my prospecting options. All the localities had been located by the BGS using computer data then verified in the field. Many of them were already well-known, such as north Wales and the Leadhills area, but they had turned up new areas. Today what I find gratifying is that the computers didn't pinpoint every gold burn or even area and completely spoil the fun of traditional prospecting. There are still burns with gold in them waiting to be discovered.

A December 1998 article in the business section of *The Press and Journal* highlighted the problem alluded to in the *Sunday Times* article. Landowners were not going to be accommodating to gold mining. There was one exception. In April 1985 a small article in *The Scotsman* pointed out that Mr. John Burton, owner of the Cononish Estate had already been in talks with Fynegold, a subsidiary of Ennex International. They, subject to Stirling District Council permission, wanted to drill for gold. By 1998, a Kildonan-born man, Dennis Macleod, who had gone from being a chemist at Dounreay in the early sixties to founding the Caledonian Mining Corporation with operations in twelve different countries, had taken over Fynegold. Dennis had always been interested in gold and other minerals, and his knowledge and research had identified eight possible mine locations in Scotland. Alas, 'Dennis MacLeod was to find that having a licence to explore was one thing, persuading landowners to allow development on their land was another' (Bob King, *Press and Journal*, 09/12/1998).

By April 2010 *The Herald* was interviewing Chris Sangster, the managing director of Scotgold. The vagaries of the price of gold had put paid to Fynegold's development of Cononish, but Scotgold had taken over and things were looking promising as the price of gold rose. What was also interesting in the article was the discussion of the geological factors in Scotland which made it "highly prospective for gold and other base metals". Chris Sangster's picture was to appear in numerous other articles.

Fact: *Over 90% of the gold processed has been mined since the Californian Gold Rush (1849).*

Almost exactly five years later *The Press and Journal* was reporting on yet another firm with high hopes of opening a mine, this time near Towie in Aberdeenshire. Again the geology was to the fore and links were being made to the similarity of the rocks around Towie to those in Northern Ireland at the "massive Curaghinalt gold deposit". A further headline said "Locals torn between jobs and preserving the landscape". Two further statements caught my eye. "The firm said the streams near the village of Towie have been well known for containing grains of "significant size" for many years…". It was news to me. A company spokesman also commented "Initially, we will cover the area with stream sediment sampling – the old school way. Then from the streams with high gold content, we will crack open any rocks we think have the potential of hosting gold". It was good to see that basic methods of prospecting were still proving useful.

"It was amazing to find real gold in Scotland" said Gill McCallum of East Kilbride being quoted (having just completed a course at Wanlockhead) in an August 2007 article in *The Sunday Times*, entitled "Treasure Highland". At least it was not the overused "There's gold in them thar hills". The piece was highlighting gold panning as one of the many outdoor activities available in Scotland. Unfortunately it had one glaring error in stating "In 2003 a six gram nugget of gold, the biggest ever found in Britain, was discovered in a river near Aberfeldy". The article was dominated by an excellent large picture showing a panner emptying his homemade gravel pump into his classifier and pan.

Aberfeldy was one of the areas where I missed out on the bonanza. I had not got round to following up rumours when I read a newspaper article headed "Gold rush blights secluded burn". It talked of how a competitor at the World Gold Panning Championships had let slip he had found a rich source. Over 50 people a month were visiting from as far afield as Wales and Cornwall and the burn was yielding gold worth thousands of pounds. Perth and Kinross Council's Ranger Service were alarmed at the activity and damage which was on an SSSI. Rumours had it that a mechanical excavator was going to be brought in to move larger boulders. The rangers were asking people to leave and most had agreed. A final quote stated "Most of the people pan gold for a hobby and they are quite sensitive to the environment".

Fact: *The word 'bonanza' comes from a Spanish word for calm sea.*

Most of the articles I collected appeared in one paper or magazine only. When I put a cutting of John Greenwood's gold panning exploits into my files, I could have taken the story from a number of papers. It was January 2012 and Grant Logan, the goldsmith from Campbeltown who had made a ring for John from his panned gold, had mentioned the story to the local press. The national press then picked up on it and before long John was being contacted by a large number of papers and eventually the BBC.

One of the best-known buyers of Scottish gold for making into wedding rings is David Milne from Drumoak, Aberdeenshire. A December 2007 *Press and Journal* article was prompted by the imminent release of shares in Scotgold. It suggested that panners were worried by the possibility that a Scottish gold mine could "bite into" the 75% premium expected for Scottish gold. Its scarcity meant that David had a healthy waiting list for products made from pure Scottish gold, despite that, he could charge £780 for a Scottish gold man's ring compared with a Chinese import at under £60. David is quoted as saying "I have heard of plans for Scottish gold mines on a number of occasions over the years, but we tend to hear of nothing more after the initial exploration stage". David's picture was in the same paper a few months later as gold prices hit record levels. He was shown melting Scottish gold for a ring for the late Charles Kennedy.

This was not the first time a Scottish jeweller had cropped up in my cuttings. In January 1998 the *Sunday Post* had an article on Ian Combe, a jeweller from Elphin in Sutherland. It was titled "Golden secret of a Highland river". With a title like that and a photo of Ian with a pan in a river, it was bound to catch my attention. Which river was providing the gold? I was not convinced the river shown was anywhere near where his gold was panned. Verbal clues were in the quotes. Ian was going panning once a week where he had been "panning a particularly rich vein in a river a good two and a half hours drive from home". He was not prepared to tell the reporter any more about the site adding, "It's secret, but I've been in touch with the owner of the land and everything's legal and above board. The river really is a terrific source of top quality Scottish gold". Ian was also looking for Scottish gemstones to make his creations.

He ended by giving the reader an insight into the variations in Scottish gold. "Kildonan gold is an orangey colour, whereas in central Scotland it has a whiter or more pale yellow hue". Was this a clue? I had my own ideas, but since he had other people panning for him it was likely he was working with some of the variety of Scottish gold colours and purity.

Another story involving the variations in the colour of gold came from one of the most interesting sources of information on Scottish gold for the layman, *The Scotsman Magazine*. Historical information is frequently repeated in the articles, but there have been little gems for the better-informed. Charlie Smart was interviewed about both river panning and competition panning. He recalled the story of the Italian panner at the 2005 British and Scottish Championships at Wanlockhead who had managed to find several hundreds of flakes of gold from a bucket of gravel containing precisely nine. Charlie explained to the writer, "Gold varies in colour and what he produced to the judges was definitely Italian gold".

Charlie Smart waits to receive a capped vial from Mary Russell at the 2014 championships in Wanlockhead. Mark Bell in the trough behind was the reigning Open British Champion that year. He lost his British title but gained the Scottish. Richard Deighton, who was later to play a major role in organising the 2017 World Championships in Moffat is in the far background.

Charlie was also involved in what was and will always remain a unique episode in Scottish history. When silversmith Michael Lloyd was commissioned to design and make the mace for the Scottish Parliament, it was Charlie he went to for Scottish gold. He made the request just before a major panning championship. Charlie and twenty other panners managed to collect twenty six grams worth up to £1,000 at the time. Charlie was quoted as saying "I'm proud to be an Englishman who is part of Scottish history".

He also had encouragement for newcomers to the pastime. "The trick is finding a good spot and working it hard, sometimes for hours at a time, then you'll find it. Scotland is full of gold – it's all over the place".

By the time Charlie was interviewed, *The Scotsman Magazine* was being printed on matt paper, but in October 1986 glossy paper was the norm. *Volume 7 number 7* left you in no doubt as to what was inside. The cover had a full-page picture of a panner using the one handed technique and the title stated "A golden gleam in the North".

The reporter had interviewed a number of people in the Kildonan area, including Malcolm McGillivray who was factor of the Suisgill Estate. When asked how he felt about the growing number of visitors he replied "We welcome them. The countryside is there for people to enjoy; it's part of their inheritance". He pointed out that certain restrictions had recently been put in place in the interests of conservation, but that it had not diminished the numbers coming. In 1985 400 licences had been issued, 350 the year before. He hastened to add that only around twenty found gold in any quantity.

Two of the successful panners added to the article. Bob Crawford was described as "a generally acknowledged guru". "What you have to bear in mind" said Bob "is that in this world there are three kinds of liar; the gardener; the angler and the gold panner: of the three, the gold panner is by far the biggest".

A further discussion brought in Johnny Marr. He was described as being "the man" when it came to finding gold. Johnny quietly reached into his pocket and without fuss laid two "small glass jars" on the table. They contained two ounces of gold worth just over £400 at that time. Johnny also showed two 22 carat wedding rings but gave the warning, "If you're lucky and know where to look, the chances are you might find a few flakes, but equally, the chances are that even if you did, you wouldn't know what you'd found and probably throw it back in the river".

The whole article was well-illustrated. One of two palm-held nuggets from the gold rush gave the impression they were into double figures in grams. Another person featured was Alistair Sangster. He sold panning equipment at the time and was holding an original metal pan. It was a broad dish with no sudden change in its curved shape.

Johnny might have been pessimistic about the beginner's ability to identify gold unaided, but it has never stopped young hopefuls getting gold fever. *The Daily Mail* on December 12th 2011 featured one such person.

I met Gareth Moore for the first time on 13th July 2011 when Leon Kirk brought him along for a film shoot. He used a tarpaulin on the banks of the Crom Allt as his base and it proved useful on the day of the shoot. We were all to act as beginners. Our tutor, Leon, gave us a lesson which appeared on French TV. Gareth was a very personable and polite young man and Colinne was immediately charmed by his good manners and helpfulness. He nonetheless was soaking up all the information and panning help I put his way out of camera. He confided that he wanted to spend the winter panning.

The article picks up the story a few months later. It described how having packed in his job in a call centre, he was living in a tepee with a wood-burning stove on which he cooked. He spent the morning gathering fuel and then went panning. He had few creature comforts and admitted he was lonely at times and would welcome someone joining him. Nonetheless he had no intentions of going back to his previous life. He thought he could make about £40 a week from gold which was just about enough to sustain his simple way of life. There were photos of Gareth panning, standing in front of his tepee with his tarpaulin over the open top and of fingers holding a vial with a little gold.

Every time I was in Tyndrum, I visited and worked with him. I was increasingly worried that his set-up was not adequate for the wet winter. Panners get wet and he was having trouble drying his gear. His tepee was too crowded for comfortable living and I gave him a tent that I'd found packed but scattered over the slopes above the Crom Allt. Walkers of the West Highland Way frequently give up at Tyndrum where they can take a train back to Glasgow. The less responsible ditch their cheap gear in the bracken. There was no way that the tent had been stowed for future use. It proved useful, but on subsequent visits I became aware that Gareth was not into a routine which would sustain him in his new life. He missed one rendezvous and thereafter I could find him still in his sleeping bag late in the morning. On one visit he was not to be found though his tepee and tent were there. On enquiring in Paddy's Bar, I was told he had taken seriously ill and had gone home. Subsequent emails elicited no reply. Eventually his camp was trashed and some gear disappeared. The mess was eventually tidied by another panner who frequented that area.

In late 2011 Barbara Copley, the BGA secretary, put out a request from a journalist who wished to speak to a gold panner. The price of gold was rising and there was growing interest in all aspects of the mineral. I agreed to contact him and he said he wanted to go panning. Warren Pole is a freelance journalist specialising in adventure assignments for newspapers, magazines and TV. I arranged for him to meet up with Leon, Jamie and me at the Crom Allt. He brought along award-winning photographer Peter Sandground. The photoshoot took place well above the railway. I remember pottering around in a pool gazing through the shuftyscope while Warren gathered information. Peter was well away but had on a powerful lens. I heard him shout get down further. I was in about two feet of water and crouched. It wasn't far enough. Eventually I was on my knees on the rock and he was still motioning for me to get my face closer to the water. Much to my surprise this was one of the photographs in Warren's article in the *Mail on Sunday* called "After the Goldrush".

Almost immediately there was an internet backlash from the angling community. As so often with internet critics, the limited information given was misinterpreted and I was accused of disturbing salmon spawning grounds. Reluctant to get into a slanging match, I did not point out that I have heard of no recorded instances of salmon laying eggs on rock far less jumping fifteen foot vertical falls to reach it. In the event, Dave Jones and Andrew Winter did respond and the storm died.

Some of the text did not go down too well either. I was a little taken aback to read "Veteran panners Alan Souter and Leon Kirk are spurred on in their efforts by escalating gold prices and the opening of Scotgold's mine at Tyndrum". Since the price of gold whether high or low has had nothing to do with why I look for it, the journalist's penchant for the dramatic gave a misleading impression. A second quote did nothing to lessen the mercenary impression being given. "People think they can strike it rich, but the reality is different. There are very few panners who even cover their costs". He went on to point out our poor returns for the day were worth "£5 at best". As so often happens, we had been pressed to give it a value.

> *Fact*: The water of Earth's oceans contains around eight times the quantity of gold mined to date; however, processing sea water to extract gold is far too expensive to make it worth while.

What I had not remembered was that I had described the numerous coalescing holes in an exposed gravel bed as "Resembling a World War 1 battlefield". This was put in such a way that it looked as though the gravel banks would remain scarred for years. Every panner knows that the first good spate would fill them all in, but the public would not. It naturally annoyed some in the panning world.

In chatting to Warren it transpired he knew my neck of the woods. He wrote articles for whisky magazines. "It would be good to do an article on whisky and gold", said Warren. The idea was discussed further and I said I would spend the winter looking for a good spot then get back to him. At this stage *The Mail on Sunday* article was not out. By October the idea had been put to a whisky magazine and they were keen on the idea. What distilleries had I mentioned? I had said there were minute amounts of gold in burns beside some famous Speyside distilleries but said that Leon would not get out of bed for the amounts. They were indeed tiny and few and far between. This was not what Warren wanted. I also suggested the Kildonan, then going to Clynelish Distillery in Brora. I had a contact there. Clynelish and Kildonan are not side by side and although tributaries of the Brora had gold, I would first have to work on permissions. The final suggestion was Glen Turret Distillery which was on a likely burn. I had found gold in some of the burns in the Crieff area and the geology was the same. I would look into it with a view to a spring meeting. On the 24th October I received the following email. "Whisky magazine are so keen on the whisky/gold idea they now want it done by the end of the month so I am trying to set something up as soon as possible". Far from having time to prospect the burn, I barely had time to research the landowners, phone them and get permissions. They all agreed.

In the event Warren booked flights for the 5th November. I had invited John Greenwood and Jamie Shepherd to join me the day before to look at the access and find a good spot, hopefully with gold. We did indeed find a few flakes in rather high water and next day Warren arrived in Glasgow and journeyed out by car with Peter Sandground the photographer. He managed to get in to my drysuit despite being almost six inches taller and duly donned hood, mask and snorkel for a photoshoot of sniping. Leon appeared with Emiel Heinneman, who had flown in from the Netherlands, and they were duly photographed in the river with Warren toasting one another with whisky even though Leon doesn't like whisky. Meanwhile Jamie, John and I tried to find a few more flakes. Was it enough for the photo of Leon with a Glenturret glass of whisky and another of gold. It was not. The gold was from a vial of Crom Allt gold that John Greenwood had brought along. With that, Warren was whisked away to try to catch his flight back to London. The article duly appeared in the January 2012 edition of *Unfiltered*, the magazine of The Scotch Malt Whisky Society. Do not believe all you see and read.

Again one statement in the article annoyed some of the panning fraternity. Leon was described as 'plundering the river base with the zeal of a one-man JCB'. If I remember correctly, Leon didn't find any gold and I'm absolutely certain Warren didn't. There was one payback for me. I got the bottle of Glenturret single malt whisky used in the photoshoot. It was enjoyed a few days later in Wanlockhead. The label is in my diary.

If John Greenwood's gold digging exploits went viral slowly, the 18 gram Mennock Water nugget found by Leon Kirk's anonymous client was almost instantly viral, thanks in the main to the fact that Leon had contacted the press through the Museum of Lead Mining in Wanlockhead. The fact that he had put a £10,000 value on it may have had something to do with the number of papers covering the story. Whether his spur of the moment valuation was correct is a moot point. Later, the finding of the even bigger Douglas Nugget, which he had agreed to release to the press for the finder, had a similar result.

Having sifted through all my cuttings I was left with one which seemed worthy of comment. It was not about gold, but as already stated, many gold seekers are interested in geology and mineralogy. The cutting is not dated, but the title is eye-catching. It was "Hunt for Scottish gems hits paydirt". Half the article was taken up with a photo of the sapphire found on Lewis by Linda Combe and cut by her husband Ian. It flagged up the fact that it was of world class quality. Also mentioned was the search for Scottish diamonds and rubies. Dalradian and Lewisian rocks were being examined for gems and places such as the Great Glen, Orkney, Ben Hope, Ardnamurchan and Lewis were of particular interest.

I know of several gold panners who have gone to Lewis in the hope of finding a sapphire. Dr. Neil Clark was successful and had a wedding necklace made for his wife-to-be, an illustration of which is in

Peter Sandground (photographer) and Warren Pole (adventure journalist) on the day they did a photoshoot for an article entitled 'In the Drink' for the whisky magazine Unfiltered, January 2012. The burn had a little gold and it flowed past a distillery. There are several others.

his book, *Scottish Gold*, page 86. Leon, having seen the necklace got excited by the sapphire. It was on display at the exhibition of Scottish gold in the Hunterian Museum of Glasgow University in the summer of 2014. He questioned Neil about the source in my presence. Neil didn't give too much away, but I wasn't surprised when Leon told me he was going to take his mother on a holiday trip to Skye and Lewis and had I any more information. I did some research and was able to direct him to the spot. On return he reported he had jumped into the sea in his dry suit at the mouth of the burn nearest the location, but found nothing.

In the 60s and 70s the Automobile Association had a magazine called *Drive*. I have a page from it with the title "There's gold in my pan", written by the broadcaster Jack de Manio. It begins, "We struck gold 145 paces to the east of the fourth telegraph pole south of the ruined cottage on the A701". There is a tempting clue for the prospector, if ever I saw one. I drove this road in 2016 but could not identify the cottage. The river involved nonetheless is unmistakable. A photograph taking up more than half the page showed Jack, pipe in mouth, pointing to a flake in a large metal pan of the type I used at the time. Jack had paid £11 5s 6d for a permit from the Crown Estate. The article also contained a map of "Where the Treasure Lies". A list of gold localities read as follows: "Mine north of Dolgelly, Merionethshire, where it occurs as quartz veins in slate. Old gold workings at Llanpumsaint, Carmarthenshire. In granite (pegmatite) vein, Bittleford, Dartmoor. Alluvial gold, Kinbrace, Sutherland; Leadhills, Lanarkshire; Glengabber Burn, St. Mary's Loch, Selkirkshire; Carnon Creek,

Cornwall; North Molton, Devon. Alluvial gold associated with tin; Lanlivery, Cornwall".

Another page from my collection features two well-known Scots whom I have mentioned previously. They were Albert Morris, who joined the *Scotsman* in 1944 and 'Fred' Adamson who at the time of the article entitled 'Coldfinger', (yes it had a C) had written *At the End of the Rainbow*. Morris describes the book as '… a highly-detailed handbook for veteran or tyro gold-seekers in Scotland'. I particularly loved the opening paragraph by the seasoned journalist which reads, 'It was a fine, soft day in John Buchan country. Rain that might have depressed even the determined spirit of the author's hero Richard Hannay fell with character-forming steadiness. In the bubbling and chilled waters of the infant Tweed, fabled in song, story and ecological statistics, I stood calf deep with George Frederick Scott Adamson (68) of Bonaly Tower, Edinburgh, looking for a small fortune in gold flecks and maybe a nugget or two that the river might part with, since it probably had plenty if one knew where to look.' Was Albert's optimism rewarded? Well he did prove it was there with five less than pinhead bits, Fred had a little more. Yet again a locality had been highlighted in the press, though this one was well beyond where I wanted to travel for a weekend.

Although, as mentioned above, my name had turned up in a couple of magazine articles, I had not as far as I know appeared in a national newspaper in connection with this hobby. That was true until August 2017. I had parked our van at one of the sites earmarked for the World Championships in Moffat. I spoke to a lady two vans along and she was invited to join Leon, Colinne and myself when we went for a meal in the Stag Hotel. She said she had seen a

Journalist Colin Freeman uses his phone to record for an article in Country Life. He went home with some good flakes. John Greenwood is feeding one of Leon Kirk's original 'Hungry Haggis' sluices.

picture of a guy with snorkel and mask looking for gold on the *Daily Express*' online site. "That must be me" I said and having seen the photo, confirmed it was one of Paul Jacob's images from the shoot done in the spring. It was not in the paper. There was a photo and interview with Leon next day, but while walking past a shop window, there was a newspaper article with a photo of me scraping away at a crack, bulb in hand. Next day almost every paper was carrying shots of one or other of us with information about the championships. Leon and I drew strange looks from the shop assistant as we paid for bundles of dailies, only one or two of which were for us. We had been asked to gather the publicity for others working on the set-up. For once, there were no adverse comments or repercussions.

The photo shoot pictures were used again almost a year later when news of a very large Scottish nugget was released to the press. Its discovery had been kept secret for nearly two years and when it was decided that a wider public should know, the finder wanted to remain anonymous. Dr Neil Clark was privy to its existence and had examined it, while Leon was asked to front the release. Since Paul Jacobs had done such a good job on our previous encounter, he was given the exclusive and did a further shoot with Leon holding the nugget. The release was further delayed as another gold treasure story reached the papers on the day it was due to go out. Leon kept me informed and a week later the story appeared on the internet and next day in the press, making the front page of one or two dailies and with good spreads inside others.

The nugget, named the Douglas Nugget, a beautiful lump of gold with quartz weighing 85.7 grams, was hailed as the largest found for 500 years. It had been found by a forty year old sniper and in order to explain the sniping method, the papers had used Paul's photos of me. Such is the position and garb of the sniper, that I was completely incognito. My local friends were totally unaware of the story as they favour the *Press and Journal*'s more local news, though a week later a satirical article by 'The Flying Pigs' picked up on the find and hinted in Doric that it might have come from the burn at Meikle Wartle, a ruse by the inhabitants designed to boost the tourist trade.

It was not the only humorous piece related to the find. *The Herald*'s cartoonist Steven Camley had a drawing of Theresa May ankle-deep in a Highland stream with shovel, classifier, and a bag for gold. In the top left hand corner was the headline, 'May:– Rest assured I have a no deal contingency plan'. The look on her face suggested she was not doing too well at finding nuggets.

With the internet and its search engines and forums and social media sites, there is an even bigger information source, but in pre-internet days press cuttings were an important source of clues as to where to look. Even today the odd article can add to your knowledge. Nonetheless, for the gold seeker, the press can be a mixed blessing. It keeps you up to date with the latest big finds, particularly commercial ones. Sometimes the information can lead to armchair prospecting and eventually some colour. On the other hand, you have to be careful what you say to a reporter or it may stir up the wrong kind of dirt.

FOUND and LOST

On my bookshelves is a small orange booklet which I bought in the USA called *Colorado's Lost Gold Mines and Buried Treasure* by Caroline Bancroft. I have read it many times and although I know the stories, I get excited reading yet again about the veins and mines and hoards of gold rich ore that have been found, but for some reason or other, the exact location has been lost.

The reader may have had a similar experience when working a profitable but relatively nondescript spot. You can be plagued with uncertainty and doubt as to its exact location when returning in a different season, or after a flood has made changes or the water level has changed. I carry a small camera in my pack and photograph my working area. If it's been some time since I worked a spot, I'll take a photocopy with me to remind me what it was like. This can serve two purposes. Firstly it helps me locate the spot and secondly it tells me of any changes. If the scene looks the same I'll set about developing it as planned, but if there are changes and I think someone else has had some fun there and been fairly thorough, I'll go to plan B. Good photos, when carefully studied at leisure at home can sometimes reveal a spot you missed.

In early December 2009 I met Leon Kirk for the first time. Despite the subzero temperatures and ice at the burn edges, we had both decided to try the Kildonan. We had reasonable returns though Leon's ability to spot larger pieces by sniping in a dry suit changed my working techniques from that day, but that's another story. My diary states "On getting home and looking at the photos, I realised I had missed the best bit". I had been stepping on a safe, ice-free spot to go from the high bank to the water without really looking closely at what I was passing over. For several days it bothered me and the more I looked at the photo, the more I felt I had missed out. Five days after getting home I set off to cover the 280 miles there and back in one day, to see if I was right. The temperature at Grantown was -9 degrees Celsius and the Kildonan was frozen right across at my spot. In the event on this occasion I was wrong and I returned with under a quarter of a gram.

How else can you mark a spot? James Hewitson prospects initially with a GPS, noting or way-marking possible good spots. I have carried a GPS to new streams too, but there are many where the valley is so deep and, or the tree cover is so extensive, that it is impossible to get a signal. I therefore favour the camera and a map for making sure I can relocate a spot. Sometimes on gold burns you will see stones lying on the bank which seem out of place. More than likely it is an aide de memoire for some panner hoping to return to a profitable spot.

The book above demonstrated it was not uncommon to lose a mine or stash of gold in the wildernesses of the USA. The chances of not finding exactly where you were on a British stream is very much smaller. It may on occasion take some time, as happened on the Suisgill one spring trip. Leon, Colinne and I were using the track to make walking easier. We were following Leon who knew, or so he said, where we were going to work. It was marked by a clump of trees, easily visible from the track. Those who have frequented the Suisgill will know there are several clumps of trees and we headed for what proved after an hour of walking up and down the banks, to be the wrong one.

Temporary loss of a digging spot is frustrating but not the end of the world. Loss of a gold mine is another matter. When I was a student I discovered King's College Library in Aberdeen had a copy of *The Mineralogy of Scotland* by M.F. Heddle. As a distraction from trying to understand molecular structure from Linus Pauling's *General Chemistry* or the more intriguing idea of Wegener's continental drift, which was not yet satisfactorily explained, I would copy pages of Heddle into a notebook. One such note from the pages on gold reads, "Undoubted sources – About half a mile NE of Loch Earnhead station about 300 feet up NW shoulder of Meall nan Uamh". I couldn't find the hill on the O.S. map. It wasn't till I spoke to Jamie Shepherd, a Lochearnhead resident, I realised the old station was not where I thought it was. Access to larger-scale maps on the internet helped locate the hillside. In November 2011 a scheduled meeting with a BBC

TV crew was cancelled as incessant rain had made the burn dangerously high. Leon, Jamie, Colinne and I headed for Balquhidder and climbed a hillside above the A84. The signs were good as the track had bits of quartz. Eventually we came to an excavation with a quartz vein near the top. Access was dangerous, but we got some samples and subjected them to scrutiny under hand lenses. There was no sign of gold but we felt this was where the mine used to be. A later talk with Jamie's partner, Alice, revealed she had done some research and found an online photograph by another researcher who claimed to have found the mine. The spot did not look familiar. Since we were there new gold-detecting equipment has become available which can detect invisible gold in a rock sample. This, plus the other information I've mentioned could help solve this lost mine's location.

Lost Scottish gold mines was a topic which turned up during Dr. Neil Clark's October 2015 talk to the BGA in Tyndrum. The mines mentioned were in Aberdeenshire. The name Crawford Moor was also thrown into the pot. In *The Wanlockhead Museum Trust Occasional Paper No1, A Gazetteer to the Metal Mines of Scotland*, Crawford-moor is one of two gold mines mentioned. The name Crawford Moor has disappeared from modern maps but the paper locates it at the Long Cleuch, Leadhills. It adds, "Gold is said to have been won here prior to 1680". Neil's book has an illustration of part of a 1635 map by Willem Blaeu showing Crawfordmuir as a large area, south of the Clyde and north of Annandale.

Who in Britain today has a gold mine to lose? The gold-seeker is far more likely to lose hard-won gold. On two separate occasions I have worked close to Chris Deyton when he has found a small nugget. As good gold panning burns go, the two burns couldn't have been much further apart in Scotland. They were the Kildonan and the Short Cleuch. The other unifying factor was that on both days it was windy. One of the virtues of the modern plastic gold pan is its lack of weight. Chris was using a blue Klondike pan. On both occasions he had put his nugget in the pan on safe flat ground, but a gust of wind flipped it, spilling the nugget. Neither was ever found again by Chris.

The Suisgill track and some of the clumps of trees in the distance. They can be used to help you find your digging spot, as long as you can distinguish one clump of trees from the other. In areas of commercial forestry timber extraction can radically alter the look of the landscape round the burn.

Weather can help you lose gold in other ways. Dave Jones (aka Rompin' dog) also tells a story of one that got away:

"I was prospecting up the Glenclach one August afternoon and the weather was awful (kept the midges at bay though) with heavy rain, the burn up and down like a yo-yo, thunder, lightning and hailstones on and off.

Well anyhow, I had been digging and pumping in the gravels for probably five or six hours pouring the stuff through the sluice when bingo, there appeared a beautiful flake of gold. Not wanting to lose this hard-won piece of gold, I carefully removed the sluice from the burn, placed it on the bank and washed the thing into my pan with a little water. A bit of swirling about later and the flake was laid bare in all its glory. It was still spitting with hailstones at this time but not too badly. So I got out my little plastic vial, carefully positioned it over the pan so that if I dropped the flake, it would pop back into the pan and as I was about to get the little feller onto my finger, a bloody hailstone hit the flake and pinged the little sod back into the burn. I had watched it sail through the air into the water and thought it was merely a case of pumping that area and getting the flake back. Not on your Nelly it was! After pumping that spot for about twenty minutes, no flake. The river fairy had reclaimed my prize and that was me done for the day. Not even a speck after that. Talk about being pigged off! Therefore I abandoned all hope and went back to the hotel in Abington, cursing and damning all the way."

I can understand Dave's frustration on a day when one good flake would have been a triumph. He joins illustrious company with his tale of woe. The ability to lose good bits of gold can happen to the best. Vince Thurkettle has been a world gold panning champion but like many, he has a story of finding and losing. Vince recounted in an earlier chapter his early willingness to follow the advice given in a book from the USA. It suggested that since gold was relatively soft, a steel pin could be used to impale bits of gold for movement to another container or spot. On finding his first nugget, Vince duly took a

John Hooper (left) about to put gold he's dabbed onto his fingertip into a vial in a tri-pan competition. Competitions can be won or lost during this manoeuvre. James Hewitson on the right has yet to expose the gold in his pan.

nappy pin and successfully impaled the nugget only to see it fly off as he lifted it from the pan. It was never seen again. Today he spends much of his time prospecting or teaching people to find gold, but naturally this is not a method he advocates.

It's one thing losing gold through your own carelessness, quite another when your companion does it for you. Norbie the Hungarian was working close to John Cathcart one day. Norbie had already taken a few pickers which he had placed safely in a pan on the bank. By now the reader should realise that gold in an open pan is never safe. John has a habit of being rather liberal in where he deposits stones he is clearing from the stream. This day was no exception and one of the 'grenades' hit Norbie's pan, catapulting his pickers into the vegetation. None was recovered.

> *The use of a stout pin can be used to differentiate between malleable gold and pyrite, or even base metal. Pyrite will resist the steel pin then crumble under increased pressure. The base metal will be far more resistant to a steel pin than gold. The pin will leave an imprint in gold.*

I was at a friend's house when the discussion got round to how my prospecting was doing. 'I've got a nugget' said my friend. On asking to see it, he returned with a lump of dark grey quartz studded with gold. It was a sample from a South African mine. The gold in it was easy to identify as it showed the classic jagged fracture and malleability, i.e. I could bend without breaking some of the bits that stuck out. It was of no interest to him and I later sold it for him. His wife then produced a gold brooch which had no hallmark. Again she wanted to sell but did not know if it was even gold. I asked for a needle and permission to press gently in a spot which would not harm its value. The tiny indent told me it was gold. She was offered a derisory price for it locally and I advised her to approach a well-known Scottish auction house who eventually sold it for over £400. The pin trick has its uses, but other than that I would not suggest you try to impale gold, especially in the field.

In most cases the gold seeker realises the loss almost immediately. In July 2010 Robbie Falconer (Robthor) was totally unaware of his 'find' until it was too late. Andrew's relation of this tale (see Good Days and Bad Days) captures the excitement of a good nugget find, but this tale has two sides. Robbie Falconer is quite philosophical about the loss. He had been finding good bits all day but admits he had not been checking his riddle and, since good Crom Allt nuggets are too big to go through a normal classifier, he must have tossed it out. Andrew was the winner on this occasion, but when he was in New Zealand working some of South Island's public fossicking areas he lost his best find. After three days of finding small amounts of fine gold he took a flake about 20mm long and 5mm wide. It was thin, but could be lifted for storage. He removed the top from his snuffer bottle but as he was putting it into the open top he had, quote, "this sinking feeling it had not gone in". He was right. It was nowhere to be seen. He pumped the whole spot to bedrock, but never found it again.

John Greenwood has a couple of salutary tales of losing hard-won gold. John had his day's takings in a pan on the bank of the Crom Allt, not far from the West Highland Way. The pan had a string attached to it. Panners often have a loop of string through the hole in a plastic pan as it aids identification if a number of people are using the same type of pan and it makes it easier than using the rim hole to hang on a wall when not in use. John's string was for some reason quite long. Some walkers approached and asked what he had found. On lifting the pan, he had not noticed he was standing on the string and the pan was jerked from his grasp spilling the entire contents into the wet grass. Try as he might, he recovered little more than one pan's worth.

His second mishap resulted from not having his snuffer bottle with him. John's back-up solution was to place what he had found in his face mask nose cover. Forgetting he had put the gold there, he decided he needed to see underwater more clearly, grabbed his mask and in putting it on…yes, you've guessed it. On this occasion the recovery percentage was higher and he got about two thirds of it back.

It shows the importance of having a secure receptacle for your finds. Besides making sure there is always a snuffer bottle in my rucksack (even when the sack is being used for other things) there are also a couple of screw top plastic vials and a pipette. You never know when you might want to give someone else a few bits.

When I first started finding gold I often used a blade of grass to pick up the bits and drop them into a plastic whisky miniature bottle backed with black tape. The sight of seeing the specks plummet to the bottom was mesmerizing. I also used the fingertip method used by those competing in panning championships. I then saw Johnny Marr keep his substantial returns in a bucket. I changed to carrying a wash-up liquid bottle to jet-wash my heavies into a bucket with a lid for travel security. This speeded up the burn-side recovery process and gave more digging time.

I can't remember when I gave up keeping recovered gold in the bucket and buying two snuffer bottles. I'm sure I saw one in a Goldspear catalogue. It certainly was a vastly improved system of gold recovery and security, but even this system is not foolproof.

I've knocked the top off one in mid-stream when taking it from my pocket. The top disappeared without my seeing it, but the gold had not tipped out. Dave Saltman, a regular competitor at the goldpanning championships at Wanlockhead, admits he once lost his snuffer bottle at Tyndrum. "Fortunately, there was little in it", he told me. John Greenwood had one chap rush downstream past him saying 'Have you seen my snuffer bottle?' and continuing without stopping for an answer. I once found an empty homemade one dangling from a string in a tree above the Crom Allt. Years later, from the description I gave him, Andrew Winter admitted it was probably his. There was nothing in it.

On a trip up the Glenimshaw I bumped into two panners with whom I had a pleasant conversation before moving further upstream. On returning later I enquired how they had got on. It hadn't gone well as one had spent a great deal of time walking up and down the bank looking for his snuffer which had floated off. He had found it several hundreds of yards away, having wasted a lot of digging time.

I have a fear of losing my snuffer, yet I frequently toss it on the bank or beach. To help me find it I have dayglow tape attached. Others have it on a lanyard around their neck. If you go to retrieve it and it is not where you think it should be, it can raise the heartbeat by a few notches.

A nugget in a bottle would seem to be safe from the kind of riverbank loss described above. Early in his gold-seeking days, Leon Kirk had come back from a trip to Kildonan with a small nugget taking pride of place in his trip finds. Like most gold seekers he was keen to show what he had found and the nugget was duly put in a glass vial. It was being handed round back in Glasgow when it was dropped, the glass shattering. Search as they might, neither Leon nor the assembled group could find the nugget in the grass. Next day Leon ran a metal detector over the area with no success. The grass area outside the Strathclyde University halls of residence still has a lone nugget waiting to be found. He has always maintained 'Plastic for the day…glass for display'. When teaching he will have the day's takings transferred from plastic snuffer bottle to a glass vial so that his client can show it to best effect to anyone interested.

This went badly wrong after a course for a group of divers based at the Wigwams near Tyndrum. The gold was divided and as one of the party, Chris, was admiring his share, he dropped the vial onto the hardcore. It smashed. I jumped forward to mark the spot and keep everyone back. A trowel and pan were fetched and I carefully trowelled the area marked. Having satisfied myself I had removed the area covered by glass, the pan was handed to Jamie. He headed off to the burn, reappearing with a grin on his face and as far as we could see, most of the gold. It was duly put in another vial and handed to an embarrassed Chris.

Chris Paterson tells a story of his collection of flakes from the Scottish Gold Panning Championships. In these speed panning competitions, the competitor is allowed to keep the flakes he has recovered, but is encouraged to hand them in for a raffle ticket and a chance to win a bigger gold prize at the end. Chris had decided to keep twelve flakes in a glass vial which he carefully stored in his jacket pocket. After the competition he met with others at the Wanlockhead Inn where his jacket was taken off. At some stage in the evening someone sat on it with the inevitable consequences for the vial. Realising what had happened, Chris decided to recover his gold later at home. Carefully opening his pocket out he was confident all he had to do was shake it into a pan and collect the bits. As the 'Compare the Market' meercat would say, 'Simple!' There is an oft quoted cliché in gold panning, 'gold is where you find it'. Wherever it was, most of it was not in his jacket pocket. He was left with three flakes.

If Chris' gold was easily come by, Judith Hill's was not. In the mid 90s, I remember a lady scraping the rock cracks in Kildonan for what she told me was to be her wedding ring. Her total was increasing very slowly. On 2nd November 1996 an article headed "Judith's 'prospects' disappear", appeared in the *Press and Journal*. The story was a gold story and I wondered at the time if it was the same person. Judith had had the unusual proposal from boyfriend Andrew who had said, 'If you get enough gold for a wedding ring, I'll marry you'. Spurred on, she spent every spare weekend panning and twice went to an Inverness goldsmith only to be told there was not enough. She finally amassed three grams to which zinc was to be added. She had showed her hard-won gold to a friend, put it in a shirt pocket, then gone shopping in Inverness. That evening she found that the small make-up sample bottle was missing and despite returning to the shops, where she had been trying on jerseys, her ring-gold was lost. She must have been heartbroken, but was determined to start again. Two days later another related article described how help had been offered by a man who had been born in Strath Kildonan and had panned there as a lad. Having worked in the chemistry side of mining and developed a process for gold extraction from gold mine waste tips, he was now head of the company he had founded, Caledonian Mining. Dennis MacLeod had exploration rights in the Kildonan area, but all his samples had been used in testing. Nonetheless he was prepared to buy gold from other panners to replace Judith's lost collection. Judith was going to go back and add to this and so the wedding was planned for the following year.

"It was you who told me not to keep my gold in glass." The voice at the other end of the phone was faint, but unmistakable. It was Victor Heatherington from Helmsdale. I had helped him find his first gold just over a year before. I wondered what was coming. I reminded him that I'd said that plastic was safer for the burn but glass was better for display'. He digressed. He'd found 9 grams. This had been built up in small amounts with regular visits since I'd helped him. He'd returned home one day and thrown his shirt in the washing machine. On going to take it out he saw a gold butterfly shaped object in the rubber door seal. It reminded him of the best bit of gold he had in his collection. It then dawned. His gold had been in a glass vial in his shirt pocket. He managed to recover three grams, but "The rest is down the river" he said. "No it's not" I responded, "It's in the first U bend in the system. Have you dismantled your machine?" "It's a communal washing machine, it's in the river", he reiterated. At this point I realised he was resigned, if not happy about the loss of two thirds of his gold and there was little sense in encouraging him to do a little machine dismantling. The conversation changed direction.

The process of cleaning gold is fraught with dangers. Those who have watched the TV programmes about mining gold in the north of America will have seen frying pans of gold being 'cooked' over a stove or fire. This dries the take prior to weighing. For years

An ounce of Kildonan gold still in a vulnerable state.

I have used a similar method on a very much smaller scale. An aluminium film cassette container from the post-war era works very well on a ceramic hob or cooker plate.

Tweezers are used to take it off for cooling. The gold is then carefully poured onto a folded piece of black gloss paper prior to pouring into a vial over a dry pan. This process in itself can be prone to mishap especially if you have an involuntary twitch.

Leon, Jamie and I have worked together on numerous occasions. Jamie is a master at cleaning up the day's findings and dividing it into to all intents and purposes, equal piles. We let him get on with it and do not quibble. It remains to decide who gets first choice. There has never been a problem. Sometimes we work close to one another but not in conjunction. In which case, finders are keepers.

On the last night of our trip to the Suisgill in 2014, Jamie had dried the gold and had it on paper when an unforeseen movement tipped the gold onto the table and carpet. That evening was spent trying to recover as much as possible, but as we left next morning, I could swear the cracks round the edge of the table had more than crumbs.

Both Jamie and George Marshall have admitted to finding gold by panning their vacuum cleaner bag contents. A hard day's prospecting on the Breckagh Burn had given me six reasonable flakes. Only five showed on transferring them to a vial. I took out the Dyson, emptied the container and began cleaning the carpet in the area where I had been working. The first pan produced a small bit of gold, but I was sure it wasn't the lost flake. I went over the area more thoroughly and went through the process again, making sure the carpet wool was well washed out. This time the flake was of the right size and morphology.

Some panners like Peter Dallas have over time developed the skills required to turn theirs and others' gold into rings and other jewellery. When Stan and Margaret Johnston's collection had built up Stan was determined to have two rings made. Having colleagues with a range of skills and equipment at their disposal, Stan decided to make his own. He set aside 10 grams of fine gold. One colleague helped him make a mould. Another provided a metal crucible, but the attempt to melt it proved unsuccessful with a normal gas Bunsen burner. It was then taken to a forge where a third colleague was in no doubt that if the crucible was put into a larger one and the forge turned up to full blast, the gold would melt, but it would take a little time. A coffee break was suggested. With mounting anticipation they returned, expecting to find the gold simmering in the small crucible. What they found was a melted crucible which on cooling had areas of gold plating. The rest of the gold had evaporated. Another attempt was suggested, but Margaret put her foot down. Another 10 grams of hard-won gold was not to be forthcoming. Thereafter the Johnstons' rings were made by professionals.

Stan Johnston's gold-plated crucible. At the time those concerned were not aware of the use of ceramic ones.

Mike Jones was another panner with more gold than most to experiment with. He was melting one and three quarter ounces of gold using a carbon arc torch in his garage. The gold was in an iron ladle held in a vice. The fierceness of the flame went right through the ladle and an ounce of gold was lost as a 'billion dots', Mike's words. Mike has a garage in which tool boxes and other items are gold-plated, well, gold-spattered at least.

In the 1980s I met a young panner from Aberdeen just above the gorge on the Kildonan. He had, earlier in the week, dug underneath a large boulder and taken a nugget estimated to be around 7g. This made him worried sick and paranoia set in. He confided that he had feared someone might break into his tent at night and steal his nugget while he slept. The solution was simple. He slept with his socks on and the nugget in his sock. Through the day he took it out again and put it in a round margarine tub into which he put all his new findings. It was still quite a common practice to carry gold around in a bucket or tub as snuffer bottles were not universally used.

Colinne and I had decided to change our spot and while walking down the bank came across a margarine tub but not a soul around. I opened it and there was the nugget and a collection of flakes. The Aberdonian had decided to move and was in the gorge when I caught up with him still blissfully unaware that he had left his gold on the bank.

Perhaps the funniest tale of gold loss was related by Peter Dallas as we walked uphill to look at the Cefn Coch gold mine during the BGA's 2015 BBQ weekend at Dolgellau. Johnny Marr was having a successful day on the Kildonan. He had found some good flakes and his gold was lying in a water-filled pan on the bank. He was joined by a friend who was out with his dog. It was hot and the dog on seeing a dish of water quenched its thirst. It was some time before Johnny noticed the dog lapping and by this time there was little water or gold left. How do you recover gold from a dog? The dog was carefully monitored and every time it left a deposit, it was collected, broken down and panned. Nothing. Determined to recover the gold, the dog was taken to a vet to see if a laxative would help recover the gold. It was then Johnny got the bad news. He would have to wait till the dog died.

If Mark Oddy wants to pan at Kildonan he has the long journey from the south coast of England to cover. He frequently did the trip without an

Members of the BGA on the Wales weekend 2015. From left to right: Tim Light, John Hooper, ?, David Saltman, ?, Mike Jones, ?, Colinne Souter, Shona Richards, Miles Light, James Linnett, Richard Deighton, Barbara Copley. Later, we panned some minute gold from a mine dump using a puddle as a water supply. No gold was lost, but that night a squall blew away the tented BBQ area.

overnight stop. In the spring of 2012 he joined a small group to work the Suisgill Burn. On day one he had the misfortune to break his pump, but I had spares and Mark worked steadily all week. Neither of us had a spectacular week, but he must have enjoyed his trip, because he was back the next year.

Mark takes up his saga as he called it: "*I was up in May on the Suisgill just short of the bridge when I poked my pump almost horizontally under a boulder forming part of a low waterfall and in the bottom of the pan came out my nugget and quite a few flakes and plenty of lead pellets. I was somewhat pleased and carefully wrapped it up in a safe place, my handkerchief. I carried on in the same place and found a bit more gold and lead but no more nuggets. I decided to call it a day and examine my find more closely back in my B&B. I collected my stuff together and climbed up the bank and while walking back I met another panner and we began chatting about our day's collecting. It was a very windy day. My eyes were watering and my nose was running, so I yanked out my handkerchief and blew my nose. Instantly I had that shock, horror moment knowing I had shaken out the nugget. Both of us then got down on all fours and started frantically looking in the grass and heather. No sign of it. The other guy, don't know his name, then went back to his rucksack and brought back a small hand held metal detector. We still failed to find it. Eventually we gave up and I returned to Helmsdale cursing my utter stupidity. Next day, I returned and started tearing up the grass and heather but to no avail.*"

Mark gave up and headed home.

"*I could picture the spot clearly, so in August I returned with my own carefully tuned detector and within a minute I had a signal. I still could not find it. We must have trodden on it back in May, but eventually I found it buried deep in the grass.*

The 'Lost and Found Nugget' (over 4g) was then wrapped and this time put in my upper shirt pocket."

Mark was lucky, but these losses should be a salutary warning to the inexperienced. I've noticed that at the beginning of gold panning courses people often ask how much gold you have or the size of the biggest bit you have found. I'm happy to talk about it, but Leon carries his collection of nuggets which he produces at the end of the day. On one course at the Mennock in 2016 he asked a lady in her eighth decade to hold out her hand. He poured his nuggets into the outstretched hand which, as she viewed the gold, began to shake uncontrollably. Realizing the danger, Leon asked her to hand them back, but one stuck in her hand then dropped, hitting Leon's boot before landing in the grass. There was a crowd around and Leon yelled "Stand still". He put the other nuggets in his vial and began a search. It was not to be seen. One of the group said he had a metal detector but it proved useless as he didn't know how to use it. Then he discovered he had locked himself out of his car. The problems had just doubled.

Another course member had a probe but this proved just as ineffective. John Greenwood, who had been the 'gravel monkey' for the day, produced a knife and started cutting the grass, but still the nugget stayed hidden. It was lost. John had a can of paint, normally used in connection with his quarry blasting activities. He sprayed an arrow on the lay-by pointing to the location. Everyone went home. Leon then phoned Davie McMichael who had been honing his gold detecting skills away from streams with some top of the range equipment. Two days later he and Leon returned, combed the spot and recovered the nugget.

John Greenwood's solution to locating a lost nugget. It worked.

These tales pale into insignificance compared to the loss of a Canadian miner I read about while writing this chapter. He had a profitable claim at the top of Goldbottom Creek in the Klondike in the eighties. His take for the limited season of about a hundred days could amount to several hundred ounces. Fearful of leaving all his gold in one container he split it into smaller ones, often tobacco tins. These he then hid. On returning home one year he realised he had left one container. Nonetheless he sold the property in 1985. Two years later he heard the next owner had discovered a shampoo bottle with 25 ounces.

I doubt if anyone has 25 ounces to lose in British panned gold, but no matter how big or small the amount lost, it is always galling if not heartbreaking to put so much time and effort, not to say expense into recovering this metal, to then go and lose it permanently.

THE PROSPECTOR

'Man cannot discover new oceans unless he has the courage to lose sight of the shore.'
Andre Gide

Jamie Shepherd has been quoted as saying 'The Crom Allt has a lot to answer for. It has limited the quest to find new, good Scottish burns.' In many ways Jamie is right. In the early teens, the Crom Allt was a magnet for panners producing good returns including nuggets for almost everybody who knew what they were doing and were prepared to work hard. As people found out how prolific the Crom Allt was, new areas were not being explored with any gusto. It is human nature to treat gold fever by going to a burn which will give a good day's return, rather than spending time on a stream which may, but more than likely may not have gold. Fortunately there are those prepared to have blank days or even enjoyable good days finding the merest signs of gold mineralisation. At the very least their activities build up the body of knowledge; at best it eventually extends the possible locations for an ever-growing number of participants.

In one chat with Vince Thurkettle he recalled how in the early days of his search for gold, people like Rex Bingham would take delight in finding a flake at a new location. This information was willingly handed on and the knowledge of British gold sources was added to. Vince had once stated in an interview that people had been looking for British gold since the Bronze Age and it was extremely unlikely that new finds could be made. Then the West Water was discovered. "How wrong could I have been", said Vince with a chuckle.

Decades ago when the only two streams I had dug were the Kildonan and the Suisgill, I remember Rex telling me he had found gold in over a hundred streams, some of which he told me, I was driving past on my way north.

The Romans, when they came to our shores, were very interested in the gold potential of Wales. It has even been suggested that their progress north followed the line of gold localities. The 16th century saw considerable prospecting and mining in central and southern Scotland. This has been well-documented in Dr. Neil Clark's book *Scottish Gold – Fruit of the Nation* and *God's Treasure House in Scotland* by the Rev. J Moir Porteous. The latter is unobtainable, but Ian Hewitson and Alf Henderson had an abstract of the pages relating to gold printed for the Goldpanners Association in 2000. Both are well worth consulting if you have any aspirations as a 21st century gold prospector.

Nowadays, when a new good burn is found, there is usually a reluctance on the part of the finder to divulge its location. This is only natural as the finder wishes to spend some time looking for the best spots and getting some recompense for his or her efforts.

This was the case at Sutters Mill in California in 1848, although there, the finder was not a prospector and at first he was not sure he had found gold. Once it had been confirmed, it was difficult to keep such a large find secret for long. Within a year it had precipitated one of the greatest mass migrations on earth. Keeping quiet about a burn is a little easier in Britain and several good finds have taken a few years to come to the attention of the greater panning fraternity. One way or another most come to light, but not necessarily in the same way.

Annually, John Hooper would encourage BGA members to bring along their collections to show others, emphasising that they did not need to say where it came from. Viewings are often done at the same time as having a drink and alcohol can loosen the tongue. This is probably what happened to start the rush to the Moness Burn. At that stage, I was not a BGA member and only heard rumours of the finds. I did not have the time to pursue a new area and by the time I did, steps were being taken to keep the public off.

Many panners mark their working spots and finds on a map. I remember seeing a 1:25000 OS map of the Leadhills area with all sorts of marks and comments on it. It had been lost by an anonymous owner, but on this occasion did not add significantly to the available knowledge from other sources. I was also given a look at Mike Jones' similar scale map of the Merioneth gold belt marked with numerous lines, crosses, circles and annotations. Mike was quite relaxed about my seeing it. Perhaps it was because he had already taken copious quantities from the area or the fact that I was unlikely to spend any time there. He even added that most of his gold came from a 200 yards stretch. Probably the real reason was that almost all the annotation related to mines and hard rock references rather than stream finds. At a meeting two years later, Mike was less forthcoming about a big Scottish nugget he had just found. I was sitting beside him at the evening lecture of the BGA Tyndrum autumn weekend in 2016. He had two vials of what he called 'light gold', from Navelgas in Spain, scene of the 2015 World Gold Panning Championships. Light gold is gold made up of very thin flakes which cover a good area but never weigh in at quite what you expect. After a brief discussion, he uttered the name of a source of light gold in Scotland known to us both. I asked him to keep quiet about it as it happens to be a favourite spot of mine, offering peace and quiet as well as a little gold. Mike then produced a small polythene bag and shook a newly-found 12 gram nugget onto the table. It was much admired both then, and later in the camp kitchen after the lecture. Where was it from? All Mike would say with a wry smile was "Breadalbane". We were at that moment sitting in Breadalbane, it's a big area. Obviously, Mike's craft at finding exceptional gold samples had not waned and he wasn't going to give a good location away for a second time. It was he who had innocently let slip he was off to Aberfeldy after the championships only to discover several others decided to follow the ace prospector.

This attitude is similar to the one I adopted on finding a promising burn. My diaries, which had hitherto been available to anyone to browse, no longer travelled with me. When not in the area, the maps were removed from my van. I even removed and hid a railway station badge on my van fridge when I went to BGA gatherings. It might have given away the rough locality of the find. Another way I covered my prospecting tracks was to omit adding good stream photos to www.geograph.org.uk, the website which has photos of most of the one kilometre squares in the GB and Northern Ireland. This secrecy gave me several years to prospect the burn more fully. Unfortunately, the hot spot I had found initially was not representative of most of the burn.

Most gold panners who explore hitherto undocumented streams will try them because they like the look of the stream. This is certainly true of Bob Sutherland when he began to explore the Angus area. Bob had been looking for a particular mineral near Banchory on a foul day. It was so wet he decided not to get out of the car, but let his dog out to chase rabbits for a while before heading back to Montrose. On the way back he stopped at Clatterin' Brig for a cup of tea. The rain had eased and on looking at the burn he noticed a number of quartz boulders. The very recognisable mineral quartz had alerted Bob to the possibility of gold. He took his pan and got two specks in the first pan. That was what alerted him to gold in the area. In the next year he had over six ounces, but was also beginning to look further afield. He settled on the West Water. The story of his initial day is contained elsewhere in this book, but his find unleashed one of the more bizarre episodes in Scottish gold mining lore. He was reluctant to tell people where he was working, but he nonetheless teased people like Eddy Bell by showing him his finds. Word was out that Bob was the man who was finding both quality pieces and gold in quantity.

In a recorded interview with Bob at the 2010 Gold Panning Championships he told me how he was followed. "Then a couple who come here" he said, "you never remember their names, if I see them I'll point them out. They followed me. They must have sat outside my b****y house for a week watching me. And on the Saturday morning I was working … and I seen this car in my mirror and every time I speeded up he speeded up, every time I slowed down he slowed down. I said, 'That b*****d's following me'. So as soon as I got into Forfar I nipped down some of the side streets and then when I got to my work I parked the car behind the sheds and I never saw him again. On the Sunday I went up the burn and I was only there twenty minutes when this b****r appeared. Of course, he told everybody and the gamekeeper was convinced I was telling everybody how to get there, so he was going to stop my permit". He then went on to tell me how he protected his permission to pan the burn. Did he point out the individuals concerned? I'm keeping that to myself for the time being. Bob was a true prospector and he left clues about other burns to those he liked. He preferred what I would call (modifying an old cliché for the panner) the 'pump it and see' technique. He went to a burn, found a good spot and dug.

I always get a little excited if I see signs of quartz in any area I go to. There is little doubt that a lot of the free gold found in Britain comes from quartz, but in over fifty years of picking up loose quartz pieces and looking at them with a hand lens, I've not found one speck. Nonetheless, some of the bits of gold I've found in the Kildonan, the Suisgill and the Crom Allt have contained quartz. In fact, hold up a vial from most burns and bring it into focus from beneath with a hand lens and you will more than likely spot some quartz attached.

A trip north or south along the A9 passes through several cuttings in the region north of Bruar. There are fine examples of quartz veins in some of them. Each time I see them I promise myself I will stop next time and run a hand lens over some of them. I

Even small burns can have gold.

have, however, looked in some of the streams nearby and found small amounts of colour. This has been confirmed by Mike Jones.

Leon Kirk has told me on a couple of occasions that he reckons Mike is the best reader of a river in Britain. Mike told me that, having decided the stream might be favourable, he doesn't jump into the nearest spot but will walk some distance, mentally noting possible gold holding spots. Only then does he pick the most likely one to test for gold. This precision was not what marked his collection when he showed it. Some vials had the names of well-known streams, but others were labelled with only the general area. Guess why?

In 2013 a group of us were sniping on the Wanlock Water. We were not doing well and James Linnett and Stuart Sutherland decided to pack up. Leon had moved from below me to above and when I eventually had had enough, I went up to him to say I was giving up too. Leon had nothing. By the time I had picked up my gear and got back up to him he was in raptures. He had a small but beautiful nugget with a lot of quartz and shaped like a croissant or quarter moon. He called it the banana nugget. Gold with its host quartz, is treasured.

Sometimes a piece of quartz in the pan will behave tardily and be reluctant to leave with the rest of the gravel. It is always worthwhile isolating it and taking it home for examination with a microscope or a powerful hand lens. As often as not the seemingly white quartz has a little bit of gold which has slowed down its movement in the pan.

Some of the bigger nuggets found in Wales have quite a bit of quartz in them and in 2013 an exhibition relating to gold and gold panning mounted in the Museum of Lead Mining at Wanlockhead showed a very fine example. This nugget wasn't from Wales, but from the Mennock Water, found by John Hooper, a past president of the British Goldpanning Association. Its discovery should be an encouragement to all beginners.

During my 2011 visit to the Cononish gold mine with the Strathfillan Saplings we saw the core room with hundreds of drilled core samples and were encouraged to take rock samples from the dump beside the mine. All the white quartz had bits of pyrite and Chris explained that the mine was full of sulphide ore. In other words, the gold is in microscopic amounts and is contained in the pyrite.

Fact:

Gold forms an amalgam with mercury. It is still used in some mining areas round the world to extract fine gold. Evidence of its use in the Kildonan Gold Rush can still be dug from the gravels of the burn.

Tips from the top:

"Remember, gold has been deposited as the burn cuts down. Sometimes it's in the rocks high up." - Bob Sutherland

"When you think you are finished, that's when you begin" - Vince Thurkettle

A small example of free gold near a quartz vein from the Cononish mine.

Chris had been employed at the mine for some time when he mentioned to one of the employees of the previous owners that he had still to see a bit of free-gold from the mine. This miner then sent him a small sliced and polished greyish core sample with thin white streaks running longitudinally and a couple of larger pinker veins at right angles. Parallel to one of these was a very thin one with a beautiful streak of gold replacing the quartz for about a fifth of its length. It was duly passed round, admired and photographed. It is another sample which could take pride of place in any mineral collector's display. Anyone who visited the mine during the long drawn out process of trying to get it fully operational would have noticed that extensive dumping of rock had built up a level used for car parking. Visitors were at liberty to pick up and take what was considered to be worthless rock, much of it white quartz with pyrite. By 2016, in an attempt to raise interest in the mine and investment cash, Scotgold R. Ltd. was processing this material, getting around 7 grams per ton.

Tips from the top:

Terry Mahoney gave me something to think about when I mentioned I had constant problems with my facemask leaking. One trapped hair will break the seal he explained. Although I shaved off my upper moustache, it had not occurred to me that my eyebrows needed the same treatment. It worked.

So did Leon Kirk. "Mr. Souter, are you blind?" he said to me one day on the Kildonan. His clarity of view in snorkelling gear showed I'd missed a small nugget in a dark hole - I ordered a dry suit, bought a mask and snorkel and have never regretted it - except in mid January.

Colinne panning near where she found the gold vein. Years later Leon Kirk covered the area carefully with a high-spec detector, but couldn't find another.

The quartz vein with visible gold described earlier, had some of the gold in the quartz, but some was on the margins and into the schist. The two best pieces, the ones that had shown in the sunlight, weighed 12.8 grams and 15.9 grams, but they are not what I would call 12 or 15 gram nuggets, as the weight of gold is probably only a gram or two in each. Heddle's map of the Tyndrum area shows several larger mineralised veins crossing the burn in the vicinity, but he doesn't mention gold, only galena and blende, i.e. lead and zinc ore. The geologists at Scotgold assure me that Halliday's vein, which has been worked high on the hill to the southwest of Tyndrum does contain visible gold.

Historically, Scotland has produced a great number of eminent geologists who spent much of their career in the field. James Hutton (1726–1797), the father of modern geology, demonstrated amongst other things that igneous rocks were once molten; a fact we take for granted today, but was revolutionary in his. Roderick Murchison (1792–1871) at one time studied the deposition of gold in the Urals in Russia and his work contributed to the discovery of gold in Australia. Hugh Miller (1802–1856) built up an enormous collection of fossils and expanded our knowledge of the Old Red Sandstone. Sir Archibald Geikie (1825–1924) was a field geologist who rose to become head of the British Geological Survey. Benjamin Peach (1842–1926) and John Horne (1848–1928) unlocked the structure of the Northwest Highlands. John Muir (1838–1914) roamed the wilds of California and Alaska furthering the knowledge of the glacial process but being better known for his conservation work which led to the formation of National Parks.

Sir Roderick Murchison took an interest in the Kildonan Gold Rush and in his capacity as Director of the Geological Survey encouraged more detailed geological investigation of the area. Two other interested parties, the Reverend Doctor J.M. Joass and Professor M. Heddle, like Murchison, believed the source of the gold was local. Of all these, it was Heddle's writing that interested me at an early stage. Matthew Heddle (1828–1897), was a professor of chemistry at St. Andrews University in the 1860s but he had a passion for minerals and mineral collection and travelled all over Scotland assembling a superior collection of minerals, particularly agates. His collection is housed in the National Museum of Scotland. He is said to have had incredible fitness and stamina and was well-known for carrying hammers up to 28 lbs. on his excursions. He also used dynamite to blast apart rocks to reveal their 'hidden treasures.' (I don't think John Muir and Matthew Heddle would have got on too well together.) His book, *The Mineralogy of Scotland* was published posthumously in 1901. The two volumes were and still are an extremely useful reference for the mineral collector. As a student, I filled a notebook of references from this work and followed up many of them in the coming years, building up a very modest collection of minerals. My notes on gold read as follows.

Undoubted sources:

Sutherland. Loose in river gravels of Helmsdale and Brora rivers – among chief localities, the Crask, head of Clyne Milton Burn, Strath Brora, the Blackwater and Kilcalmkill, Strath Ullie, Helmsdale River, Kildonan Burn, Suisgill, Allt Ceann a Phris.

Perth. At Corriebuie mine 1½ miles from Tomnadashan in limestone overlying mica slate at 2,000ft. veins carrying quartz and argentiferous galena also gold –About ½ mile NE of Loch Earnhead station about 300 ft. up NW shoulder of Meall nan Uamh.

Leadhills in Lanark. Windgate Burn, Green Louther.

Doubted sources:

Perth Glen Quaich at Turrich near Amulree – only one nugget found

Improbable:

Durness Sutherland – Tyndrum Perth

The last location is interesting and surprising bearing in mind that the Crom Allt runs through Tyndrum. It could be explained by the fact that although lead mining was important in the area little heed was given to other minerals. Traces of gold had been found at the Earl of Breadalbane's other prospects at Lochearnhead and Corriebuie. No one had bothered to prospect the Tyndrum area for gold, or at least if they had, they were keeping very quiet about any finds. Heddle wrote over four pages on gold in Scotland and they are well worth reading even though we know considerably more than he did about where Scottish gold is to be found. Copies of his book are rare, but its pages can be accessed on line on 'The California Digital Library'.

I didn't copy every reference. For example, Heddle refers to gold on Unst. In the early 90s I went to Unst without remembering the reference, taking only a fishing rod. Years later a friend living on the island sent me a Shetland Island Council report entitled "Precious-metal distribution in Shetland: refinement of targets for gold exploration". While there, I had been staying in the area with the highest concentrations of gold (over 1,000 parts per billion). One of the maps indicated thirteen sites on Unst where gold had been found in panned concentrates. I had missed an opportunity which was unlikely to arise again. The maps in the centre of the report nonetheless alerted me to the idea of pathfinder minerals. The varying concentrations of arsenic on Unst coincide very closely to those of gold.

Heddle also mentioned the Kildonan and the Suisgill under his "localities which have been doubted – SUTHERLAND. Kildonan Burn, in an ochry quartz vein about 1 mile from mouth. In visible specks in the granite of Suisgill". Here he was looking in detail at suggested sources and not the alluvial gold. Ochry refers to having iron oxide in it (ochre) and would certainly have a reddish staining in the quartz. I haven't looked for this vein, but I do have some small pieces of gold covered in a red dull coat. Gold coming from hardpack can also be covered in a dark iron deposit and may not be instantly identified as gold. Getting to know the slow speed of movement in the pan helps to avoid throwing away good samples. The Argyll Lochy is a river where this can be the case.

My early diary also contains my painting of what I called an 'oddity' in the text. By that I meant at the time that it was not gold alone. The painting is nonetheless labelled as gold in quartz. Far from being odd, it is exactly where you should expect a great deal of gold to originate. Today, I would photograph such a find down a microscope.

This is my painting of gold in quartz as seen under a microscope. It went into my diary. I now have photographic equipment to do the same.

Heddle's second reference regarding gold in granite, is less likely to be true. Gold certainly gets wedged very tightly into cracks and under flakes in the Suisgill granite, appearing to be part of the matrix, but whether it is in the initial mix is doubtful. In granitic areas it is more likely to have come from secondary enrichment along with quartz veins. Having said that, I have found very small amounts of gold round some granite areas such as the Etive and Ben Rinnes granitic intrusions.

There are other ways of narrowing down the search area without leaving your living room. One of the best guides to gold localities is Dr R.J. Chapman of Leeds University's *Gold Mineralization in Scotland: a rough guide for goldpanners*. The acknowledgements include help from a number of panners, several of them well-known, including Vince Thurkettle, Chris and Richard Deyton and Mark Gregory who provided samples for analysis. The pamphlet deals with how these samples were analysed and what it tells the scientist about where the gold originated in the earth's crust. What is of much more interest to the panner is the section headed 'Some important panning localities'. Eight localities are named, The Calliachar Burn and Moness River; The Crom Allt Burn; The Glengaber Burn; Leadhills-Wanlockhead; The Sutherland Goldfields; Angus Glens; The Ochil Hills and The Lammermuir Hills. Many of these areas have already been mentioned in this book and the access to others is not freely available to the panner. The illustrations of gold from various localities have a key which names some of the burns from which the samples came. For years a large plastic wall version of this graced the wall just outside the toilets at the Museum of Lead Mining in Wanlockhead.

> **Fact**: You don't have to go far to find gold - our bodies contain about 0.2 milligrams of it, mostly in our blood.

The BGS also has a gold pamphlet freely available on the internet. Called *Minerals in Britain – Past production… Future potential – Gold*, it was published as a glossy six page pamphlet but can also be downloaded from the internet. With these two documents and internet access, the armchair prospector can discover a large number of burns and areas. In fact they open up so many, that it would take years of panning to cover them all. Let me give you an example.

The *Minerals in Britain* pamphlet has a map of all the gold deposits in Britain. It classifies them by type, e.g. Strata bound, Mesothermal, Porphyry and Alluvial to name four. For the most part, these categories need not concern the armchair prospector. Having said that, not all deposits give free gold in the local streams. The armchair prospector wanting to find gold for the first time would head to the areas showing alluvial gold i.e. gold in water-deposited gravels. These include Helmsdale and Leadhills-Wanlockhead in Scotland and Mawddach Estuary in Wales; plus a couple in Cornwall. The Mawddach Estuary reference is a little misleading as, although it is thought that there could be workable commercial deposits there, it is the River Mawddach itself that gives panners Welsh gold. But if you want to find new gold burns then you have to go beyond the well-known areas. One of the names that is shown on the map is Flowerdale Forest.

Using Google and searching 'Flowerdale Forest + gold', one of the four sites that turns up is the BGS pamphlet in pdf form. Another is entitled *Mineral exploration Lewisian supracrustal and basic rocks*. This file is definitely not for the non-geologist, but even within the 70 odd pages, there are little hints as to whether the gold here is worth going looking for. On page 57 there is mention of a concentration of 4082 ppb au. If this was from a sediment sample I would be adding this to my 'to visit' list. Unfortunately it was from a small rock sample of a sulphide ore including copper and zinc. Added to that, there was little mention of native gold. Checking the location on an O.S. map showed it was well away from the road and the nearest river was not where the sample had been taken. As a result, this is not a locality I have visited. This is despite having found a reference to one person who was convinced he had found specks of gold.

A third link is about copper, but even it has references to gold since the two minerals are often associated. The final link is to the Minerals Reconnaissance Programme, (MRP) 141 to 146. There are numerous references to possible gold deposits in the various synopses including Flowerdale Forest, but number 144 which is about the British Permian and Triassic red beds has two statements which catch the eye. The first is "gold (is) found to be widespread and locally abundant in alluvial sediment from the Mauchline Basin of red beds". The second states "streams draining the Thornhill Basin were also auriferous". This, as far as panning is concerned, is far more interesting and requires further research. This is where you go back to Google and search "Mauchline Basin + gold". A further pdf can be found. It takes some time to sift through it, but pages 24 and 25 are of interest to the panner. Here I add a word of warning. In Rob Chapman's pamphlet he has the following statement. "The gold is usually very fine grained, but can exhibit spectacular crystalline forms. This gold type is probably more widespread than currently realised – but unlikely to make you rich!" In other words, if you are looking for large returns, these are not the streams to prospect, but if you want to prove it is there, you have a wide range to choose from. The same sort of armchair prospecting can be done for the Thornhill Basin adding another group of streams.

I used the Rob Chapman information and the Geological Survey Ten Mile Map (North Sheet – Third Edition (solid) 1979) to find three gold streams in Banffshire and Moray where the red rocks of the Middle Red Sandstone lie on top of much older impervious Dalradian rocks. Looking for the rock boundaries on the geological map and using the OS 1:50000 Elgin and Dufftown map (No.28), I made up a list of streams that were close to the geological boundaries. I reduced it to four where the contour interval or waterfall symbol showed me there was likely to be rock close to the surface. Only one was a failure. Of the rest, one gave only a few minute specks, the second gave a little more, with a few flakes up to two mm. From the third I can take 30 to 40 specks and flakes per day. The biggest bit is 0.1 grams and finding such a bit is like finding a nugget in the best burns. The aim is to have a vial with two grams. The gold is associated with barytes, hematite, magnetite and ilmenite in such quantities that it can take a long time to isolate what little gold there is. Since I'm several hours from a good gold burn, having found these local sources means I can relieve gold fever with a day trip. This came about as a result of armchair prospecting and was all the more gratifying because none of the lists of gold burns that I have seen, has a Moray or Banffshire entry.

Mining companies, while doing exploration, publish a lot of data much of it in map form. Some of this can be accessed on the internet. Scotgold Resources Limited (SRL) were very encouraging to the amateur prospector, providing speakers for BGA gatherings and allowing access to their licence area. A form was produced which allowed the panner to put in a small sample for analysis with details of the place and physical nature of the deposit. When I was directed to submit a couple of samples to Rob Chapman at Leeds University, I was told he couldn't take any more. Mining companies get clues from what is found in streams, but it is the quantity of gold in bedrock that they wish to know. The few grams that amateurs take are peanuts compared with the hundreds of thousands of ounces that might make mining viable. Nonetheless their annual reports show where there are gold anomalies and where rock samples have given promising grades of gold expressed in grams per ton (g/t Au).

Even before SRL was on the scene, Ennex International had explored the Tyndrum to Dalmally area and produced a map, a copy of which was reproduced in *At The End Of The Rainbow* p.91. The units used here were oz. Au/ton, but the locations were shown as dots on an otherwise blank map. Fortunately both Tyndrum and Dalmally were marked and by rescaling the map and superimposing it on an OS 1:250000 scale I was able to locate a number of gold burns, most of them now well-frequented.

The BGS's website Minerals UK had for some years an interesting download related to the Department of Trade and Industry's (DTI) Minerals Programme Publication No.12. It related to metalliferous mineral potential in the Northern Highlands of Scotland. At the bottom of the first page of the online work was a map showing three red circles, numerous patches of red in two tones and 28 blue stars in lines and clusters. The red showed areas of gold anomalies and the stars, gold in panned samples. The problem was there was no grid reference nor any names. At first this seemed a useless map for the armchair prospector. Fortunately there were two small coastal outlines at opposite corners. One had enough indentation to identify it precisely, the other had few features. It did have a couple of stars at the coast. If I assumed it was a major river, the Helmsdale, I could again square it up with 10 km squares. This done, I gave each star a grid reference then started checking them on the 1:50000 map. Was I surprised by the place names that turned up? Anyone who has read about the Kildonan Gold Rush will know the names Suisgill, Kildonan, Torrish, Allt Breac and others. But there were others that I had suspected might be gold burns. Several were in more remote spots and the BGS had used their manpower and authority to confirm them. What of the gold anomaly areas? I looked at one or two and could have added some more blue stars to their map.

Another exploration firm which spent quite a bit of capital on looking for Scottish gold was Alba Mineral Resources PLC. Unfortunately for them, after identifying numerous gold areas in Perthshire, they could not raise the funding to keep going in Scotland and let their licences lapse. Nonetheless, their annual reports make interesting reading on a dark evening and were still available on the internet until recently.

Soon after I had met Leon he began mentioning a large book he thought I would find interesting. In reality he hoped it would point the way to a good burn. When he lent it me, it turned out to be the BGS's Regional Geochemical Atlas of Argyll. It had numerous maps of the analysis of thousands of stream sediment samples. There was one map which did have Au in one or two places, but they were already well-known. What I was looking for was new areas. I concentrated on the maps of three pathfinder minerals, antimony, arsenic and bismuth. Bismuth was already known to be the contaminant in Crom Allt gold which lowered its purity and made it brittle. The correlation between arsenic and gold deposits is well-documented. My notes stated "All major As anomalies Kilmelford, Comrie, Tyndrum and the Etive intrusion all reported to have gold enrichment". The Etive intrusion was not known to me as a potential gold locality at the time but the geological map showed me where I might find a good burn. At the same time Scotgold's 2011 report indicated one burn near the Oban to Ballachuilish road where I might have success. I found gold within 50 metres of a car park, but further superficial exploration proved disappointing. Other areas were noted but I didn't get round to prospecting half of them and those I did were not worth a return visit. The Loch Earn

Scotland's first 21st century gold mine. This is what professional prospecting aims to find.

area showed up as a potentially good area, so I mentioned it to Jamie Shepherd who lives there. Two years later when I asked if he had looked at the burns with the highest values, he hadn't. The returns in the Crom Allt, although further away, were too good I guess. For a couple of years, the atlas was frequently left open on the dining table, surrounded by notebooks and clear overlays. This did not go down too well in some circles, but was tolerated. Eventually there was a request to look through the mass of information from an up and coming panner. Colinne was quite pleased when I gave it back to Leon. In my notes, I have more burns to look at than I will ever prospect.

When Andrew Winter heard I was interested in looking beyond the well-known burns, he offered me a map which he said showed fault lines. The thinking behind prospecting burns in the vicinity of faults is that research has suggested that an earthquake caused by fault movement may instantaneously deposit gold. The detailed process is not of importance to the prospector, but it is thought that the deep hydrothermal fluids with dissolved gold have a rapid pressure drop which allows the gold to be released. Further work in Australia and other areas has resulted in statements like, "small scale fault systems in the earth's crust have a strong correlation with the location of gold". There are a number of gold localities near Scotland's Highland boundary fault and in fact if all known gold localities are plotted on a map with the major fault lines, there is a very strong correlation. Some lesser faults have also proved to be gold-bearing localities, the Tyndrum fault being a prime example.

Fact: During an earthquake, veins of gold can be deposited almost instantly from hydrothermal liquids.

Andrew also mentioned he was looking at lamprophyres in his prospecting. Lamprophyres are igneous rocks that occur as dykes and small inclusions. I had suspected that some of my finds were coming from areas with a preponderance of what the Geological Survey Ten Mile Map North Sheet third edition showed as thin, relatively short, lines of rock. Rhyolite, trachyte, porphyrite and lamprophyre were all classed as intrusive. In the 5th edition map, their naming was changed to simply sills and dykes and a geological dictionary had the first two classed as extrusive rocks. Ignoring the confusion, I asked a question on the BGS website as to whether there was any correlation between these rocks and gold. The response was non-committal and not a little disappointing. After that I began to find online articles which did link lamprophyres to gold. In one, the map of Scotland had the thin lines of rocks that had interested me labelled as lamprophyres.

I began to follow up this research in Perthshire and on a trip to Crieff I pointed out to Colinne a number of very small streams which cut across the dykes on the map. I boldly stated that they should have gold, but did not stop. I did test one on the return journey but in little over an hour I was unable to find any sign. I had thought that I was too far down this stream, but hoped any gold had carried. I was disappointed and thought that my theory was incorrect until a few months later I came across an archived mining report of the area, and lo and behold each and every burn I had suspected, had given up gold in panned samples. The burn I had tried had no gold where I panned but did have further upstream near the intrusion. The Central Highlands have numerous areas with these dykes and sills, many in remoter areas where the fit and adventurous prospector might discover a new burn of note.

Getting down to this sort of geological detail is more for the professional, but may interest a few amateurs. The BGS 1:50000 Series maps can also be worth a look in some areas. The Leadhills sheet 15E solid edition has not only the rock types and the fault lines, it also shows quartz veins and the metalliferous veins which gave rise to the mining in this area.

When Greenore Gold began announcing they were finding good signs of mineable deposits near Towie in Dalradian rocks in Aberdeenshire. I was keen to see them at work, especially if they were sampling burns. Unfortunately, I didn't make contact, but it did flag up the importance of Scotland's Dalradian rocks as a source of gold.

Before I head to a new burn, I now do a little more homework online. www.fishpal.com has a section which gives river levels. This doesn't mean I'm prospecting rivers with migratory fish, but it helps to know what the last rain has done in a river basin. With experience you get to know indicator streams and levels. The Scottish Environmental Protection Agency (SEPA) www.sepa.org.uk also has water level gauges in an even-wider range of river basins. There are eighteen pages of recording points listed alphabetically by station. This means you have to know the station name and look at a number of different pages for one river basin. The main river is named in the second column under the heading 'catchment'. The third column is the actual stream name. To use this site quickly it pays to have a list of the pages and station names for the streams you are interested in. The other problem with the site is finding the pages. On the home page select environment at the top, then click on water, then water levels in the box on the left. The data are accessed by clicking on water data in the box on the right.

Two other websites can be of use. Google Earth and Google Street can help show possible access and parking spots. You may also get an impression of the stream environment i.e. is it gravel or rock. www.geograph.org.uk may have photographs of the stream you want information about. It is not always the case as many more photos are taken from roads and paths, often ignoring the main stream in a kilometre square. This site can also give you access to OS maps including the 1:50000 and 1:25000.

I have a small display box which holds four half gram vials. They contain the accumulated takings from the best four streams that year. They are discretely labelled so that the onlooker cannot see the names. Two of them I'm not prepared to divulge, yet. At the end of the season I empty them into the total collection for that stream. On the lid of the box is a cartoon gold miner holding a big lump of gold and leaping into the air. We all like to celebrate a new or good find and may well taunt other panners with a glimpse of it or a hint of the weight taken. On the one hand we get pleasure from showing off our skill or luck, on the other we know we are not going to have many more good days at the locality if everybody knows. In 2015 and 2016 all I would say about two vials was they were my secret streams. One was secret because I wanted to protect the landowner who had given me his permission to prospect. Word of a good stream can result in a gold rush, with some panners barely respecting the landowner's wishes. This happened in 2017 on another Perthshire stream I had visited and as a result, everyone was banned. The other stream is known to a few panners, but reported totals were not high. I thought there could be better returns if I had enough time to explore more fully.

The BGA newsletter is a source of good gold stories. Kit Andrews, the editor, displayed exactly the same ploys in his article in the 2016 summer edition. The description of his prospecting was such that it would excite the keen prospector or even the gold-digger who just wanted to amass some gold. The stream was not named and the location was vague, but there were enough geographical clues for the armchair prospector to make an attempt to locate it.

The armchair prospecting I've described is time-consuming but it is an interesting aspect for long dark evenings. It can be simplified by someone looking for a burn to pan, by studying the lists of gold burns in G.F.S. Adamson's book *At the end of the Rainbow*. There are also lists on the internet. At the end of the day, research will probably fuel your gold fever and help you choose the time and place. It may even point you towards a good burn, but it doesn't put gold in a vial; for that you need to get out as often as possible and process as much gravel as possible.

The sort of rock structure that traps gold.

AWAY GAMES

'The great and recurrent question about abroad is, is it worth getting there?'
Attributed to Dame Rose Macaulay

Colorado

Verbier, Val d'Isere, St. Anton, Courmayeur; these are not places familiar to most of the gold panning fraternity, but mention them to those that like 'white gold', in other words snow and they'll tell you they are among the top ski resorts in Europe. Through the eighties and early nineties I'd skied these and many more Continental resorts. In only one Austrian resort did I come across any evidence of gold panning. In the early nineties I turned my attention to North America and booked a trip to Steamboat Springs, Colorado. Enquiries about local minerals were a disappointment. In a country where there had been numerous gold rushes, there was no mention of gold. They did have coal. Much as I liked Steamboat Springs, the next trip was to Breckenridge in the same state. I couldn't miss the signs that I was in gold territory this time. One of the restaurants was in a floating gold dredge, another was called 'The Prospector' and I had a drink in the Gold Pan Saloon. Even some of the T shirts had a panner. In winter, there was no hope of digging, as over a metre of snow had fallen. What I could do was buy books and maps and make enquiries.

The best-looking title was *Summit – A Gold Rush History of Summit County, Colorado* by Mary Ellen Gilliland. This was studied night after night and maps were pored over. Summer return flights to Denver were booked for two. We found gold in a number of creeks and rivers but never in any quantity. With two days left we were persisting in

A prospector I met on a Gold Prospectors of Colorado claim. Here, methods inappropriate for Britain can be used. He will, nonetheless, have to restore this ground when he is finished digging (see www.gpoc.club/about-us/).

French Gulch to the east of town. The literature had described gunfights over the riches here and I had hoped to do well. We were working in old gold dredge tailings, using a trowel to dig out small amounts of clay from between the cobbles. Most people had buckets to carry half an hour's digging to the creek. We followed suit. It was slow panning because the clay was hard to break down and why it had passed through the dredge sluices without releasing all the gold. We began to have good pans. Now that we knew how and where, we returned the next year and concentrated on the same area. We also continued to explore other creeks. Nowhere beat French Gulch and we got a picker or two in most pans. We had a profitable trip and after another ski season were keen to return.

The Klondike

Despite having had several goldrushes, Colorado is not the best-known of North America's gold areas. California, the Klondike, and Alaska were all higher profile events. Alf Henderson opted for a trip to the Klondike, where the name Henderson is part of the history. Robert Henderson was a determined Yukon Territory prospector with Scottish roots. He had been unsuccessful on the Pelly River, a tributary of the Yukon, but was grubstaked for two years to prospect the Indian River another east bank tributary. He found fine gold but was not satisfied. On one occasion he fell on a branch which went through his calf, and he lay in his bivouac for two weeks before continuing his quest. He climbed a large domed mountain to the north from which several creeks radiated. Choosing the one to the north he struck paying gold, establishing a discovery claim on what he called Gold Bottom. Had he headed west he would have been in Eldorado Creek or Rabbit Creek, later to be called Bonanza Creek. Henderson missed a second chance of becoming a very rich man through his openly racist prejudice to local native prospectors, two of whom were in the party of three which discovered the gold in Rabbit Creek. Having made it plain he did not want them staking on Gold Bottom, their white companion failed to tell Henderson of the even richer creek they had discovered, despite a promise to do so. Henderson's fortunes went from bad to worse and he left the Klondike penniless.

George Alfred Henderson, known as Alf in the gold panning fraternity was a teacher bitten by the gold bug. He too found the allure of the North American goldfields too much to resist and chose to go to the Klondike on four occasions. Alf's access to the Yukon varied. Twice he hired a car in Whitehorse, once he drove a campervan purchased in Edmonton then sold it at the end of the trip. He even travelled up the Top of the World Highway from Vancouver.

I've mentioned his full name because he had to use it to swear on a Bible that he had claimed an area on Treasure Hill, Lascelles Creek, Hunter Creek in the Yukon. The cost was only $25 but taking a claim was a serious move for a person from the UK. Alf had to have a box number as an address and renewal depended on doing 200 hours of work. Road work didn't count, but toilet provision did. Alf did not try to establish a claim on his first trip. He used it to make friends locally as well as finding gold. It is rare that a visitor to these goldfields will take much gold on a first trip without help from someone who knows the area. Two local claim owners allowed Alf to work the drains where sluicing had spewed the waste material after separating out the gold (the sluicing process is not always efficient and big sluices can lose as much as 30% of the gold when not set correctly). Alf found a rich one and took five and a half ounces between 11am and 3pm. He admitted it, but unfortunately he had to leave for home next day. Barbara Copley was not leaving and took eleven ounces panning in a tub.

When he returned to Canada, other locals helped him find a suitable bench claim which he held for three years. Being a bench claim there was little water and bucketing it for panning was a major drawback. In fact he sometimes had to use puddle water such was the problem. Alf also jokingly said there was another closer source of water if things got too difficult. Was he joking? The claim was rich. When Alf relinquished it, a New Zealander used more mechanised methods to take out 900 ounces. At one stage Bob Sutherland was keen to pack in his job and go with Alf. Worried that the setup might have changed, Alf didn't take Bob. Bob later had his own claim, but Alf admitted that he felt Bob never fully forgave him.

At the BGA 2014 autumn get-together in Tyndrum, Alf produced a wide vial containing a considerable weight of Yukon gold. Although he kept a good sample he did, however, defray some of his expenses by selling even more for tourist lockets and to put into bags of paydirt.

Taking out a claim in North America is a serious undertaking, especially if you live on the other side of the Atlantic. With a limited season in the northern gold rich areas and the cost of getting there, the would-be gold miner from Britain has to think seriously about the commitment required. Nonetheless going to the claims office and checking on what has not been claimed is a sensible move if you want to go prospecting. At the BGA 2016 autumn get together in Tyndrum, Mike Jones told me that he too had prospected an area of bench deposit. He added that there was good gold in it and it had not been claimed. I wasn't sure whether he was casting a fly in the hope that I would rise, or just teasing. Either way, with the World Gold Panning Championships in Moffat in under a year, the next summer season was not free for either of us.

Alf also told me briefly about a trip to Africa where he came across a gold-bearing river with enticing cracks across the bedrock. Before he was able to work the area, his wife had an accident and had to be taken to hospital. The trip progressed and Alf found himself near the same river but further downstream. His anticipation rose as he approached, but his hopes were again dashed. The river had crocodiles at this point.

New Zealand

If the gold digger has not made a contact in a foreign gold field before travelling, you hope to make good links on your first visit, something I singularly failed to do in Colorado. I did not make the same mistake before heading to New Zealand. Use of the internet can help to make contacts and I found a site run by Allan Dunford, aka Snuffy, who was offering trips to Otago burns with his dredges. At least I would have a mentor for part of the trip. His reply to my approach began "Thanks for the mail – it's always interesting to hear from gold people". Other early contacts included an email to the Department of Conservation (DOC) regarding the cleaning of my gear. New Zealand is particularly vigilant regarding the import of foreign species. I was assured that if I followed the guidelines my pans etc. would not be confiscated. As instructed, I presented myself in the queue for those importing equipment. I had read the likely consequences of fines or imprisonment for those who tried to flout the rules. I was contemplating this while sinking my hands into my jacket pockets as I waited my turn. I suddenly froze and a shiver ran up to the back of my neck.

A month before I had been on Ben Rinnes looking at path work and the reseeding of badly-eroded areas. I had collected heather and grass seed to propagate plants for transplanting. I had forgotten a polybag, so I stuffed them in my jacket pockets. I was wearing that jacket and had not fully emptied the pockets. On approaching the desk, I explained my problem to the quarantine officer and waited for the admonition. He was charm itself. A vacuum cleaner awaited and a little later my jacket and our gear was returned with a smile and good luck wishes.

There was plenty of evidence of foreign species causing problems during the trip, particularly the extensive areas of broom, but it was Didymosphenia geminata (Didymo) that was concerning DOC at the time. The internet carried pictures of a grey wool-like substance wrapped on a stick. It had been imported from foreign rivers, probably on wet angling gear. Various gold rivers were infected and the gear sterilization precautions involved made it prohibitive to pan a variety of streams.

New Creek, New Zealand - not the easiest stream to get a day trip return from!

Well into the trip I was hanging over a bridge on the Buller, one of the infected streams. It was a clear stream and I could see large trout in the greenish water. A car drew up and two attractive uniformed ladies with aviator sun specs approached. I was asked if I was going to enter the water and I explained I was just watching the trout. I told them I was aware of Didymo (there were so many notices you could not be unaware) but I had yet to see it. I was told to wait a little longer and it would probably appear. It would look like strips of toilet paper. Within ten minutes the whitish Didymo came drifting down. It was just as they had described.

This alerted me to the need for care when panning in foreign areas. It highlights the responsibility gold diggers have to the environment. The care doesn't end at the pre-trip preparations. In Scotland almost all our large streams and river have a population of migratory fish, either salmon or sea trout. There is a parasite in foreign rivers which could have dire consequences for a fish population which has already suffered from attacks from diverse agents. Gyrodactylus Salaris can be spread by wet gear that has not been properly treated before use. Every time I booked a week on the Suisgill, the estate owners would send a sheet with information on how to treat gear which had been abroad in the last sixth months.

Access to 'gold fossicking areas' in New Zealand is easy and well-documented. I say easy, in that no permits are required. Physically getting to them can be a little more difficult, especially if you are driving a normal hire car. We had opted to use a car and motels rather than hire a motorhome. Some of the places we had to drive to get near the panning areas would have been impossible for a motorhome. A rugged 4x4 would have been more appropriate. At one location I had to spend an hour extricating our car when the ruts became so deep the wheels became airborne. I was grateful for having bought a long-handled shovel. Next day we took the advice that the road was really only suited to 4x4 vehicles, left the car at the first ford and walked five kilometres uphill in baking heat to try New Creek. We were finding small amounts of gold at every location we tried. Most of it comprised of small thin flakes. It occurs to me that since almost all the streams had courses littered with large cobbles and boulders, the gold travelling with them is beaten flat.

In Shamrock Creek we found grittier bits. Here the bedrock was a grey clay. Colinne took our biggest bit and best pan from a small pothole I had pointed out which had already given me a few flakes. On prising a flat stone from the base, she discovered it protected a rich spot. So often potholes in Scottish streams are ground smooth and are goldless – or is it I've just come along too late? Only once on the Erochty did a pothole prove to me the presence of gold, but not a lot.

It was at this location we were robbed. I was standing a little way from the car when I heard a loud shout that our car was being targeted. The warning came from a large motorhome made from a converted bus. On looking round Colinne said "It's got our lunch bag". I reassured her lunch was in my daypack, but Colinne was insistent. Only then did I realise a Weka, a sturdy flightless bird that had been wandering around the car, was heading off trailing a familiar polybag. I gave chase and one half sandwich spilled out. I was sprinting at a reasonable speed, but I could not keep up. It entered the bush and search as I might I could find no signs of bird, bag or sandwiches. Lunch that day was a shared half tomato and meat sandwich.

You don't have to go abroad to be robbed by the locals. Norbie the Hungarian was working on a tributary of the Mennock and had yet to settle on a spot. He left his pack in full view and was amused when at some distance from it, he could see sheep gathering round it. Unconcerned, he set about digging, returning to his rucksack only when hunger set in. It was then he realised what had been attracting the sheep. His food lay scattered around, each item having been bitten at least once, but not entirely devoured. Not relishing sheep-chewed sandwiches, a starving Norbie packed up and headed for the car.

Finding our contact's town was relatively easy. Having stopped at a hotel, we were told to turn left after the shoe fence. We had already seen a bra fence near Cardrona, but were still surprised by the variety of footwear decorating the fence on the way to Naseby. Allan Dunford was not in, but after trying to find the grave of my gold mining ancestor in Dunedin we were to return for two days dredging with a 5 inch dredge.

The first burn, the Hogburn, did not seem a likely spot for panning, far less sluicing. It was totally dry. Allan excused himself and disappeared in his 4x4, reappearing a little later. Thirty minutes later a trickle of water appeared and in no time the burn was in flood. Allan had opened a sluice.

The dredge worked well, with only the occasional need to hammer the intake pipe to loosen jamming stones. My back was playing up so I stood watching in the rain while Colinne dug gravel and panned. I got steadily colder and was on the verge of hypothermia when we packed up. On emerging from the tree canopy the reason for my feeling cold was very evident. The hills around had a fresh covering of snow to just above the level of the plain. The previous day we had been in shirt sleeves. New Zealand was behaving a bit like Scotland, but back home I would have checked the forecast. Allan did not seem to notice the cold. The dredge had taken 1.5 grams of very fine gold to Colinne's 0.1 g.

Dredging in Muddy Creek, Otago. The dredge processed plenty of gravel and gave a good return, but unlike panning, we did not know how much we were finding till the end of the day.

Next day we tried Muddy Creek, another source of very fine gold. In assisting with the dredge, I missed the excitement of seeing the first signs of colour in the pan and then the full revelation, good or bad. Before long we were digging our own holes. This time the sluice to pan ratio was reduced. Allan had 2.9g, we had 0.4g, Colinne taking by far the bulk of it.

The hospitality from Allan and his wife Denise was second to none and although the outings were amongst the most memorable gold days, I was not fired up to buy a dredge. Had I lived in New Zealand perhaps I would have thought differently. I had hoped we would meet again, but when he did come to Scotland we couldn't meet up. While preparing this chapter I learned that Allan had died in 2012.

Andrew Winter left New Zealand in 1997 but makes fairly frequent trips back home. In early 2012 he decided to include some gold digging on his trip. He targeted the public fossicking areas on South Island's Aorere River and Lyall Creek. Hoping to snipe, he found both streams difficult for a one day trip as he "was confronted with large gravel banks and fast running water". He did rather better on a third stream on his second day there, but only after getting some clues from a local.

California

In 2013 Malcolm Thomas the BGA president and Chris Paterson went to America to compete in the California State Open and US National Gold Panning championships. Malcolm, a regular visitor, had to defend his title as champion in the open event.

After, they teamed up with a friend to find some new gold in the interior. Having driven for an hour from the last habitation along a logging road, their guide Steve's 4x4 ahead came to a halt and refused to go any further. They all got into Malcolm's borrowed vehicle and headed through the undergrowth, bashing their way past bushes until they arrived at an old mining camp which was to be their base for the next three days.

They had come well-prepared with a tent, eight tins of beans (they were on offer – buy three get one free) bread and meat and cheese slices. What else does the pioneer prospector need? Well, a kettle would have been a good idea for a start, but after emptying their first tin of beans, they had a container to boil water. On hearing he was sleeping on the ground, Malcolm baulked saying it was too stony. Steve slept in the car, Malcolm in the cab and Chris had the tent to himself. He was happy with this till he heard that bears visited the area. There was messy proof on the ground not far away. As a result every little nighttime noise raised his heart rate.

They were there to work on one of three claims in the area called Cold Creek, Freezing Creek and Ice Creek – Chris tried bathing in one and thought they were well named. Nonetheless in the Californian summer it was more like dry creek or boulder creek. Two boulders in particular caught their eye. Steve assured them that once they had removed them there would be nuggets beneath. With this incentive they set about the mammoth boulders. First they cleared round them finding a little small gold. The thought of the nuggets beneath spurred them on. They then created a channel along which to drag the rocks. Using a chain around the boulder and a hand winch attached to a stout tree, with the aid of pinch bar the tons of rock were removed exposing the bedrock. The only trouble was, it had taken two and a half days. With half a day to get the nuggets they set about with renewed vigour with the expectation that every crevice they opened would be the lucky one. By the end of day three they had nothing. It was decided to prolong the trip but by day five they were down to two rock-hard crusts and a slice of bread. They had also run out of fresh water and were having to boil river water. Chris says he thought Malcolm was showing signs of starving so he was given the final meal. Did their endeavours have a happy ending? They did live to tell the tale, but they did not find the anticipated nuggets.

Alaska

The lady beside us on the plane said we would have a very good landing. What gave her this insight? Her husband was the pilot and he had honed his skills landing navy jets on carriers. Anchorage runway looked enormous and we landed like thistledown. The pilot's wife had flown in from New York to get a supply of fresh fish and would return next day. We were there for a month and we hoped to find a little gold. Next day I picked up a 4x4 pickup, visited Walmart for a shovel and spent several hours in Alaska Mining and Diving perusing and buying gold panning paraphernalia and books not obtainable in the UK. One was entitled *Where to Prospect for Gold in Alaska without Getting Shot*. The previous night I had spoken to a Yupik Eskimo (his term) who had a ring with pickers round it. He got them on his tribal lands where he said his father was the head man. He offered to take me there and showed me a remote spot on the map. Had I been on my own I might have chanced it, but I wasn't going to drag Colinne on a risky venture.

My planned first port of call was Hope, where in 1893 there had been a small gold rush to Resurrection Creek. When at last I approached the main street I wondered if it was any less risky than Yupik territory. The tin shacks and undulating potholed street depicted on the internet and which I had thought was the preserved historic town, was in fact the main street. We booked into Bear Creek Lodge and found our log cabin. It had no internal toilet. The toilets were a hundred yards away and bears roamed the area at night. We found a new use for gold pans. The owners gave me permission to pan the creek passing through the site. I was able to show them it had gold, a fact of which they were not aware. Several miners and claim owners ate breakfast in the lodge restaurant and I got chatting. I was eventually introduced to Kelly (I was told you don't ask a second name) who agreed to take us to his claim.

Next morning we arrived at the appointed rendezvous and followed Kelly's pickup up an ever-narrowing steep track. He stopped by a quad bike. "One of you will have to start by walking", he said. I motioned Colinne onto the back seat and off she sped holding on to a stranger she had only just been introduced to. She has frequently recalled since that she suddenly thought "What am I doing? I know nothing about this guy". We needn't have worried as Kelly soon returned to pick me up.

Crossing Cottonwood Creek in the Dutch Hills, Alaska, with Mount Denali behind. While panning, the motor was kept running to alert bears of our presence.

We were given free rein on his claim and got good pans. We asked where his storage container was, but he indicated he was not interested in dust and even pickers. He was looking for nuggets weighing ounces and he had recently found one. Every day on arrival, Kelly moved 100 stones to clear the next area. We did the same as a thank you for the freedom to dig.

He taught me a new technique which he called 'boot scooping'. He carefully moved the gravel back and forth with his feet so that the gold sank, then cleared off the surface layer before processing the gold-rich gravel. Sometimes this was done above smooth bedrock and slowly the current washed away the gravel revealing the gold which he sucked up with a bulb. He obviously used the technique a lot as his waders all leaked. It was made plain that we were welcome to come back each year, but we have never returned.

While Kelly was a chance acquaintance, I had made contact with a claim holder for the second part of our trip. Dennis Garrett of the Alaska Freegold Company gave me permission to have a look at the Blue Ribbon mine in the Dutch Hills. I was able to view a trommel for the first time and view the range of working and rusting equipment we now, due to TV, take for granted will litter gold claims in this part of the world. I could not work on the claim as it contained an area called The Potato Patch, so-called because the nuggets tended to be found in clusters along lines. I was, however, welcome to try in Cottonwood Creek nearby. There was also a state recreational area at the end of the Petersville road and my research had shown there were pickers to be had there. We tried both of these for a few hours over two days and got a little gold. Just getting there each day was an adventure, but worth it for the views of Mount Denali to the north which formed an impressive white backdrop.

On returning to our accommodation after the second day beyond Petersville we heard that a nugget had been found lying on the dirt road. New material had been spread to regrade the surface and it must have been taken from an area with gold. Next day we were leaving the area but stopped to use the metal detector at the roadside where the new material was obvious. There were no glints against the sun and all I got was the signal from hot rocks.

During our trip we tried a number of state recreational areas and one 'mine' open for an entry fee to the digging public. All of them produced gold, but one day visits were not the way to amass any weight. Most places would have benefitted from the use of a sluice and at the mine people near me were using a crowbar to open the bedrock well away from the stream. I ended up panning in a puddle to expose some good flakes. Although I had visited the claims office at the start of the trip and had come away with some maps, it was primarily to avoid getting into trouble on claimed ground.

The other goldfield country which is popular with Brits is Australia. It's a country I had started to investigate as a result of an elderly relation coming back with gold in a vial. She had been to Mudgee, New South Wales and her hosts had taken her to a local creek to find colour. They would be happy to show me around. I was off to a good start. I had a contact, but in the end I did not follow it up, though I did look further into gold prospecting tours with a guide. There was a range to choose from.

My relation's gold had been taken by wet panning, but Australia has large arid areas where the metal detector is the tool of choice. My interest in a guided tour was a result of a meeting with Paul. I did not know Paul's surname, so I called him Ozzi Paul when referring to him to my friends. Paul had appeared at the panning championships in his large converted Mercedes blue and white van. Round his neck was a large Australian nugget. At the time he was going to Australia for our winter. He and a friend would spend profitable prospecting weeks in the bush. It was nonetheless clear from what Paul said that this was not an environment for the tourist gold seeker.

The combination of tourism with gold panning is fun, but if you want large amounts of gold, settle for one area, get to know the locals and give yourself a lot of weeks and maybe even a return visit.

WHAT TO DO WITH IT?

'There's probably about a hundred pounds (worth) in there, but it isn't the money. There are easier ways of earning £100, but finding gold in a little stream like that, is such a thrill.'

Vince Thurkettle

Almost every non-panner who is shown a vial of gold will ask 'How much is it worth?' I used to respond by saying "It doesn't pay for the beers". The financial value was and always will be of little interest. Having said that, when the price of gold goes up, it does raise a smile. I have never panned for financial gain. Even if I had, I doubt if I could have made a profit from it. Mike Jones confirmed my thinking. Having amassed one of the best collections of British gold ever seen, he found his family wasn't interested in it as a collection. Rather than deprive the Welsh nation of an interesting and valuable asset, he negotiated a deal with the Museum in Cardiff. On reflection he wondered what his rate of 'pay' had been. Having tallied the hours undertaken he discovered it was £1.70 per hour. Even someone who manages to find over an ounce a year (there are few) is not going to survive on the proceeds. Nonetheless there is a demand for United Kingdom gold and the only source until a gold mine is opened or in the case of Wales reopened, is from the panner. I use the word in its broadest sense. Apart from selling it, what else can the panner do with what he or she finds.

You could give it away. This is exactly what several members of the British Gold Panning Association did. The Scotland Act of 1998 resulted in Donald Dewar's famous words, 'There shall be a Scottish Parliament'. This triggered events which trickled down to Scottish gold burns. Michael Lloyd, the Galloway silversmith, was given the commission to design and make a mace for the parliament and he had just three months to do it. His vision included the use of Scottish gold and for that he approached Charlie Smart. Given the short timescale it was fortunate that he spoke to Charlie just before the 1999 panning championships. Charlie and twenty other panners were able to collect 26 grams of Scottish gold which now forms the gold centrepiece on the silver mace head. On it are inscribed the words 'wisdom, justice, compassion, integrity'.

Apart from being involved in such a rare opportunity, there are two main alternatives. The first is to set it out as a display collection. The second is to turn it into jewellery.

It is hardly surprising that Colleen Jones has a beautiful necklace made from what Mike found. Anyone who has met Colleen at the gold panning championships cannot have failed to notice and admire the quality nuggets hanging round her neck. There may be other necklaces around, but I haven't seen or heard of one. Far more common is a gold ring. I say more common, but they are still rare. I have recounted Stan Johnston's tale of lost gold in trying to make a ring but Stan eventually had two rings made from his gold. They were made with pure Kildonan gold and were so soft that rather than wear them, they were kept in a display box.

Peter Dallas learned the skills of lapidary and making gold rings by being in the Scottish Minerals and Lapidary Club. He is one of the few panners who makes rings not only for his family but is also trusted with other people's gold, including John and Jean Fisher. I first came across John on a very wet day in Tyndrum when I passed with visible panning gear and triggered his rapid exit from his car. We were heading for roughly the same spot, but as I recall, I did very badly that day and he did well.

Mike Hobden (Frogescue) is another panner who was not content to keep all his gold as a collection. He wanted to make his own rings. Knowing how brittle the high bismuth content Tyndrum gold is, he researched cupellation, a technique for removing base metals from noble metals such as gold and silver. An online search for the process will ilicit videos, one of which has dire warnings that the process could result in severe injury or even death. Mike's description of his experiments were at one and the same time terrifying and funny. Since the cupel or mould was formerly made from burnt antler or fish spine or other form of calcium carbonate, Mike had tried bones and sea shells. My notes of the conversation contain the words, 'smelly mess, shrapnel, exploding (mussels) and boom'. In the end he found limpets made the best cupels.

Colinne and I have known one another since we first went to secondary school. At university in the sixties, I promised her a Scottish gold ring. It was around 40 years later that I eventually got round to having one made. By this time Colinne had been panning with me for years and was very proficient. Some of the gold was hers. It wasn't a lack of gold that had delayed the decision, but two other factors altogether. I realised I found gold in its raw form far more alluring than anything made from it. The more serious obstacle was who to trust. Could we be sure that it was our gold being used? I was surfing the net looking for a reference to a big nugget found at Tyndrum, when Leon Kirk's name popped up. I eventually landed up on the jewellery designer Shirley Paris's site where Leon was described as "Scotland's foremost panner" and the provider of nuggets for Scottish gold rings. At this stage I didn't know Leon, but I liked the design of the ring illustrated. We decided our gold would be in safe hands. Having made an appointment we called in to Shirley's shop in Larkhall and immediately took to her. A lot of the gold I had selected for the ring was from the Crom Allt. It was obviously much darker than the Kildonan gold in the other vial. I warned her about the bismuth in Tyndrum gold and how it could make the gold brittle. Having sorted out the design and size, we left 15 grams of gold (5 grams will make a band). When we went to pick up the ring Shirley said there had been a problem with the Crom Allt gold. She had had to take it to a friend who could take it to a higher temperature than she could, but he had burned off the bismuth and the gold had worked very well. We were delighted with the ring. It

weighed 9 grams, was hallmarked in Edinburgh and was 750 or 18 carat. We were handed back four grams of unmelted gold plus a packet of filings and melted bits. Some years later, Colinne was speaking to Bob Sutherland who informed her that "Tyndrum gold doesn't make jewellery" whereupon, holding out her ringed finger, she put Bob right.

Having one Scottish gold ring but two daughters to inherit leaves a problem. The answer was to have a second ring made. This time I wanted the provenance to be a little different. I went through my vials and took gold from 50 Scottish burns. Again I entrusted the work to Shirley, but rather than repeat the design, I asked her how she might represent river pebbles. We then agreed on the final form.

Vince Thurkettle's daughter Daisy did not need her father's gold. "I've panned for gold ever since I was about three years old and could stand in the shallows and not get blown over" she told the *Telegraph*. She had all the gold she had ever panned. When she was about to marry Martin Roper "there seemed to be nothing more perfect and romantic than using it for (their) wedding rings." The press release showed Daisy and Martin in Wales collecting the last gold for the rings. Martin had never panned before and Daisy had taken him to the River Wenn near Dollgellau. They were also involved when the gold was melted, turned into rings and polished.

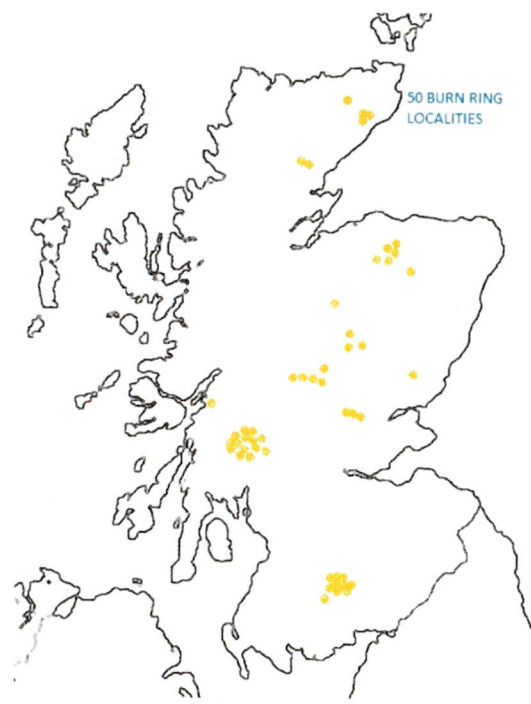

Some localities are well-known, others are not regular haunts of panners. Before trying to work out the less well-known ones, bear in mind that from some streams, all I had was a few minute specks after a day's work. There are hundreds of other burns if you want to do the same.

I was not aware of the intentions of the panner who used to waken me driving past my van in the early morning on the track to the Crom Allt. It was only later that I met John Greenwood. His quest to find enough gold for an engagement ring was highlighted by the press after his girlfriend, Morag Shearer said "yes" to his proposal. She was aware that he had gold fever, as he would disappear early on weekends and often worked fourteen or fifteen hour days. His hard work paid off and with a little help, initially from Jim Anderson, he amassed 34 grams in the first few months. He approached goldsmith Grant Logan from Campbeltown to have a unique engagement ring made. The words Scottish Gold were inscribed on the inside. He took Morag out for a pizza but drove to Gleneagles Hotel and proposed beside the Christmas tree. Grant told

the local press and before long the nationals were carrying the story. Then TV got in on the act and John was to be seen panning in Perthshire. John had more than enough gold to have two wedding rings made and by 2015 when he and Morag appeared on BBC's *The One Show*, he had added an eternity ring. As the *Daily Mail* headline read on 6th January 2012, "DIY gold ring panned out perfectly".

The BBC obviously liked the idea of young romantic men looking for the raw materials for an engagement ring. Adam McIntosh, who deals in precious stones from his tree house base near Elie in Fife, had featured before on TV. When he decided to propose, he wanted Scottish gold, Scottish 'ruby' which is pyrope garnet and smoky quartz. The last two can be found near his house. He would need help to find the gold. John Greenwood was contacted to help. He contacted Leon and it was the latter who featured with Adam in finding some gold in the Mennock. Close inspection of the film showed John in the background. Did Adam's girlfriend accept his proposal of marriage? Clare Balding, the sports commentator, was handed an envelope on *The One Show* with the answer. To her obvious relief and delight, on opening it, the answer was "yes".

I have already mentioned Ozzi Paul's nugget on a chain but others with large Australian nuggets like those Geraint Evans bought, keep them wrapped up. On rare occasions they are taken out to show anyone interested.

Making something from goldfield-finds is nothing new. From time to time gold artefacts turn up for auction on the internet. They often take the form of a pick and shovel, sometimes with a pail or a pan. Alternatively a nugget is put on a tie clasp or pin. I have one with a nugget with character, which has been handed down in the family and I've always assumed it was South African as one grandfather spent a lot of time there. I also have an antipodean goldfield memento. On my first morning in New Zealand, I went for a wander close to the hostel in central Auckland and on glancing into a jeweller's window noticed a little gold shovel. It was duly purchased. For years it sat in a box at home till Colinne said she would like to wear it. Many of the pick and shovels have a pin, but to weld a pin onto it would spoil it so I suggested having a clasp made to attach to the little loop on the back of the shaft. This would not spoil the original item but allow use of a substantial chain. These gold field artefacts could become a collecting passion, if rather an expensive one, but I've yet to see one made from home-panned gold. I know one goldsmith who has made pan badges in silver and she would have no trouble making a gold version.

The most common methods of displaying a gold collection is in loose vials. More and more often, the panner is putting the vials in a display case. The latter can house an entire collection or sometimes is limited to the best few streams. Until recently, dedicated vial display cases were not readily available in the UK. I bought a hardwood box from

Here is about five ounces from six different streams. Your gold collection is valuable; if you wish to show it off, make sure it is secure.

the Far East, having first worked out the measurements that would suit 3.5 ml (1 oz) vials. The first one went missing en route, but the seller sent a second without a quibble. I duly sectioned it, painting the inside matt black. While I was doing this Leon was working on one for selling. The vials were cradled in blue velvet hollows which really set off the gold beautifully. He wanted a picture for the front. He had a photograph of an old panner which I have also seen in various drawn forms. I drew yet another, changing the water's edge rocks to make it different from the ones I had seen. It worked out well, though for months while he worked on his sluices, made more pumps and changed his website, they were not generally available online.

Perhaps the nicest homemade display case I've seen was made by George McIvor from Brora. He showed it on an online forum I once looked at. It used what looked like red velvet, not blue. Black and purple velvet would also make good backgrounds.

If a good, well-documented collection has no inheritor it is reasonable to sell to a jeweller or museum. At home there have been jocular moments as to what happens to my modest hoard. My daughters are on record as saying they would have it melted down for jewellery. I have retorted that I'd rather scatter it in the burns of source. Then again, I could take it with me, after all, what are the detectorists of the future going to look for when all the Roman, Viking and Saxon hoards have been found.

I will be frank, I do not like the digging of British gold purely for sale. Unfortunately it's on the increase. It encourages a more ruthless element of society which does not mind causing serious environmental damage. It makes it difficult for recreational panners to find any friends in the authorities which control land and water. Britain does not have the scale of gold-bearing gravels seen in North America, Australia and the African and South American continents. Already some burns are practically worked out. It is little wonder that many now keep their new sources a secret, myself included.

The secrecy of others I respect, but I'm not enamoured of the current financial motive by individuals. What particularly irks me is the use of the internet to sell tiny flakes of Scottish gold, often shown with a blurry picture making them look larger than they are, there being no ruler scale and ascribed as having a very dubious carat. Small flakes have been described as nuggets. Some have been described as coming from the Kildonan burn where the permission to pan stipulates that finds should not be sold. The Reeves in their tenure of the Suisgill estate were good enough to allow panning with rules which meant anyone could enjoy the excitement of seeing gold appear in the pan. Some put this in jeopardy by openly selling their finds. One follower of internet gold sales mentioned that he had noticed a seller of Scottish gold had purchased a quantity of Alaskan gold. Are you thinking what my source suspected? The fact that Scotgold's Cononish mine had in 2016 started to produce small amounts of gold with an impeccable provenance might eventually help to reduce this openly monetary trend.

I wrote at the beginning of my diary, that I often wondered if I would have headed to a goldrush. With the opinions I have just expressed it does not seem likely that I was made of the right stuff. There were no hobby gold diggers in a goldrush. Circumstances are different in the UK. The Kildonan Goldrush did not last long and hundreds of 20th century Brits and even Continentals have been grateful that most Scottish streams were not totally stripped of gold.

Fact: The earliest worked gold artefacts are from Varna, Bulgaria and date to around 4500 BC.

TAILINGS

The dictionary defines tailings as "the rejected or washed away portion of an ore". I have already given an instance of where tailings were worth excavating and processing. At the end of a day, it is also worthwhile trying a pan-load of what has been shed from your pan or sluice. Finding nothing in this check is good news. What follows is a potpourri of gold related tales and observations which did not fit into previous chapters or got missed.

Meeting Nature Face to Face

'All my life through, the new sights of nature made me rejoice like a child.'
from *Pierre Curie* by Marie Curie (1867–1934).

By its very nature, the quest for gold takes the practitioner into the wilder parts of the landscape. The opportunities to encounter wildlife are greatly enhanced. What panner has not seen a small trout swimming into the hole they are digging? It will often hang around for some time as gravel and potential food is disturbed. It is an even more pleasurable experience if you are sniping and the fish is only inches from your nose. Gazing at the bottom through a facemask also often reveals the ability of the stonefly nymph to handle currents you would think would wash it away.

Encounters of an aquatic kind come in many forms. Twice I have found a toad swimming in my pan after I have been pumping gravel. The unfortunate creatures were released none the worse for their adventure. One of the delights of being at Baile an Or or walking up the Suisgill track in the spring is to hear the croaking and see heads peering at you as you walk past. On one summer visit to the Crom Allt a common frog took up residence on Colinne's yellow jacket and was reluctant to leave. Gold retrieval is not so important that I can't stop to enjoy these meetings and on occasion, photograph them.

One of the most memorable experiences happened late one evening on the Kildonan. It was in the days when I lived in a two person mountain tent when at Baile an Or. After a summer day's digging, I often felt grubby and a bath was in order. It consisted of swimming in the falls pool. I had hauled out onto the rock at the tail of the pool and was sitting, naked, slowly drying. In the half light I was aware of a movement to my right and on turning saw an otter on the rock. It looked at me, then seemingly unperturbed by my presence, it slipped into the pool from which I had just emerged. In the poor light I could nonetheless see a line of bubbles heading for the white water. I sat there motionless hoping to see a head appear, but within minutes it climbed out at the same spot I had, wandered over the outcrop, looked at me and disappeared downstream. It was a bit early in the season for salmon to be in that pool but presumably there was no harm in looking.

Later in the season, after heavy rain, it is not uncommon to see salmon running in the Kildonan and Suisgill. In November 2016 I had booked a week on the Suisgill and was praying the low water would last. For two days I worked at clearing coarse gravel off a rock shelf. There was some gold, but I was hoping that even better lay in any cracks or traps in the rock. On day three heavy overnight rain had all the burns in full spate and it was impossible to continue at that spot. With two days left and the two others in the party on their way home, I went back. The first thing I noticed was a gently waving shadow showing against the gravel in the pool above. It had not been there at the beginning of the week. Closer inspection showed a slight fork at the tail end. A salmon was resting. Suddenly it was joined by another fish barely a quarter its size. At one stage there were several. I slipped into the spot I'd cleared (it wasn't spawning territory) and was immediately awarded with a small picker. A dark shape appeared to my left. The channel I had cleared was being used as an easy way up. The fish took fright, whacked me on the head and disappeared in the peaty water. Lying horizontally in about a foot and a half of water, I settled to clearing a trap. The next encounter made me jump and unsettled me. The fish came up under my body, hovered six inches from my nose and decided to rest there. Here was an opportunity for a photo with my underwater camera. As I tried to drift down to get out, it too headed for its date on the spawning grounds. The reason for some fish using my channel, when the main one was deeper and wider was a bit of a mystery and decidedly disconcerting. The rain was on again and whereas the fish had no problem swimming, I was having difficulty holding station. I gave up sniping and pumped a little gravel.

This was not the first time I had seen a fish use a gold digger's channel to head upstream. Once when working with the sluice near the water's edge, the gravel further out was giving up gold and black iron. A trout darted up the incline and with a final wiggle near the top riffles, disappeared into the water beyond. What had happened to the gold? The iron had been disturbed to expose more gold than had been showing previously but, apart from a few flakes moving down, there didn't seem to be any loss.

It is not only fauna that has delighted me, flora too can be a temporary diversion. Primula Vulgaris or the common primrose is found on many gold stream banks in the spring. I still remember my delight at seeing Saxifraga Oppositifolia, the Early Purple Saxifrage, growing over a concrete lintel lying in the Crom Allt below the railway bridge. Further downstream and in many other wet rock and gravel areas the Yellow Saxifrage (Saxifraga Aizoides) will make me pause. So many less-frequented gold areas are in relatively inaccessible gorges. Here the flora can be limited but in the sunless areas there is a variety of mosses, many creating beautiful patterns which pass your nose as you clamber up the steep sides. I have been known to stop and take photos in precarious spots.

Toadstools on Rannoch Moor.

A toad at Kildonan.

Primroses by my garden stream - it has gold.

Rescued 'trout' from the Glenimshaw.

The Camera Never Lies

'What I like about photographs is that they capture a moment that's gone forever, impossible to reproduce.'

Karl Lagerfeld (1933–2019)

My early gold diary has one or two sketches, the odd glossy photo and a few scanned copies of photos. I had started photography as a student, taking 35mm slides. These were useful to my career, but I very rarely carried my precious camera while panning. Graduation to an SLR with a telephoto lens meant I was even less likely to carry a camera when digging, but from March 2003 my visual record of things of interest increased due to the purchase of my first digital camera. It was light and reasonably compact and also had a fast autofocus. Suddenly I was coming back from panning trips with action shots and a record of each burn. These I could peruse at leisure and plan future forays. In fact I have often enlarged them to show things I hadn't noticed when at the river. The fact that with digital photography you can check immediately that you have a clear image and retake if necessary, is important. Over the next few years, I reduced the size of the camera I carried then looked for one to take underwater shots. The first attempt was to buy a casing for the camera I was using at the time. Changing the settings with diving gloves and mask on proved a major problem. The results were not great. A dedicated underwater camera improved the shots considerably and I began to have good images of what I was uncovering. On a couple of occasions the photographs showed that my eyesight was not as good as it used to be and that I had left good flakes in a crack. One particular shot of a spot on the Suisgill plagued me for eight months. The gold was still there when I went back in the autumn. The photographing process can take time. I don't keep the camera in the water beside me as the underwater environment of the gravel-moving sniper would soon cripple a camera. As far as I'm concerned, the memories recorded far outweigh the loss of gold retrieval time.

Suisgill gold on granite. Gold on rock can be isolated by wafting the gravel away. In a fast current in a rock pocket, the gold 'dances', but is not likely to move far.

Halt! Who goes there?

'Conscience is the inner voice that warns us somebody may be looking.'
H.L. Mencken (1880–1956)

In a file of gold related articles I have a permit from the Suisgill Estate numbered 376. The dates are 31/5/ ? to 2/6/ ? It is not as early as the one illustrated in GFS Adamson's book, *At the End of the Rainbow*. His was number 7 in 1976. Since mine would have been for my annual jaunt on a long weekend, I think it must have been in the late 70s or early 80s. It has a pre-printed signature M.C. MacGillivray, factor, Kildonan Farm, Helmsdale. I picked it up from an outer room on the farmhouse when I arrived late (the road north in these days was far longer than it is today) on a Friday night. It states that I could pan on the Kildonan, but the permission for the Suisgill was by then scored out. Both these burns were at the heart of the 1868 goldrush, but in late 1869 the Duke of Sutherland refused to issue any more licences having come under pressure from his tenant farmers and anglers. By the end of 1869 the Kildonan Goldrush was over. In the mid 20th century, the Kildonan was still a rich gold burn. Having a permit meant I could dig without fear of challenge, as long as I stuck to the rules. Three lines at the bottom of the card state, 'Please ensure litter is removed, dogs are in strict control and that there is no digging on the banks, use of sluices or mechanical devices, nor pollution of the water'. The rules as stated in the Baile an Or shelter today are basically the same and do not prevent the recovery of gold, while protecting both the natural and farming environment. The other two permitted panning areas are in the Leadhills area. The Buccleuch Estate and the Hopetoun estate have similar rules, the former banning panning from the spawning grounds of the Mennock from the beginning of October to the end of May.

But what of other streams? With time, I became aware of numerous other streams, access to which was not always as straightforward as the three mentioned above. Several good burns are strictly 'no go' areas now, often as a result of landowners resenting the mess left or the excavation of banks. The Borland Burn in the Ochils was one such. A deep hole had been excavated into gold-bearing gravels away from the burn. Unfortunately, a pregnant cow had fallen in and broken a leg, resulting in the need to have it put down. The gold diggers concerned did have permission to dig at this spot, but is it surprising the farmer's attitude changed? Ascertaining who to contact to see if there is gold in a stream is often difficult, particularly if you live far from the area. It helps to speak to locals and knock on doors and explain what you want to do. I have had permission from numerous landowners, but also from gamekeepers, farmers and on one occasion from a shepherd. The person on an estate who is working daily around the area you wish to prospect is often the more appropriate person to approach first. He or she will know the feelings of the landowner or be able to direct you to who has the authority to say yes or no. I have frequently referred to the website www.whoownsscotland.co.uk, for addresses, but the face to face approach has been more fruitful.

One thing that I've noticed is that if a stream has gold and the initial prospector has had permission to pan, others assume that the permission has been freely granted to everyone. This may not necessarily be the case and can later lead to access problems.

Home from Home

'A crowded campervan is better than an empty castle.'
Anon

Some like Charlie Smart are lucky enough to live immediately beside a gold burn. Most of us have to travel hundreds of miles to get to a good burn. The logistics of this make for a long and strenuous day trip. Gold panning should be about relaxation and fun, so a longer trip reduces the rush. This means accommodation is required. There are still one or two who use a tent. They are few and far between. I gave up using a tent when Colinne decided to learn to pan. This was precipitated by my decision to go panning in Colorado. She put her foot down about using the two man Vango Force 10 tent bought to explore Iceland in the 70s. Instead we enjoyed Mrs. Polson's B&B in Helmsdale with its "high cholesterol special breakfast" (her words) and the occasional weekend in the Dornoch Castle Hotel (when they had an offer on). The journey home from Kildonan was in a Vauxhall Calibra Coupé and we would often struggle out at the Little Chef at Tomatin, laughing at our inability to straighten up.

When I retired I often looked at the secondhand car adverts with the intention of buying a small van for solo prospecting trips. My first ever vehicle, a Mini van, was sometimes my home for a night and I thought I could kit out a van specifically for panning. Dan Haddow worked out of a van, as did 'Ozzi' Paul whose large Mercedes boasted a welcome gas stove. John Wikinson's van had a workshop, but unfortunately it was stolen in England and used for criminal purposes. Mike Fisher added cupboards to his van for his trips north. Karl Heinz Riegler, from Germany, and his wife Brigitta would spend three weeks every summer at Wanlockhead then Kildonan using a combination of a people carrier and tent. Colinne was obviously not happy with my idea, but knowing my feelings about caravans and towing, she came up with an alternative. "Why don't we get a campervan". I was stunned, but quickly arranged visits to various dealers. We landed up with an

A gold meet at Baile an Or.

Autosleeper Duetto which served us well for ten years. Many panners have a campervan or motorhome. Bob Sutherland's sale of gold gave him enough to buy a very comfortable van in his later years. Dave Saltman managed perfectly adequately in a less spacious VW van conversion but later appeared in a larger home. Others like Malcolm Thomas have large vehicles with all mod cons. Leon Kirk's motorhomes have been used as a panners' meeting room, office and shop. I have yet to come across a panner with a really large rig in Britain, probably because it is important to be able to drive on narrower Highland roads and pull off into relatively small spaces. The ability to unhitch a caravan and use the car is what Mike and Colleen Jones favour. Mike even has a mountain bike on his 4x4's roof to get to places his vehicle can't. Some like 'Klondike' Dave Grey, leave a caravan in Helmsdale, while John Wilkinson, after losing his van, got permission to park his caravan in Leadhills.

There are one or two exponents who have not succumbed for one reason or other to using a mobile home of some sort. Dave Jones and Andrew Winter book accommodation for their longer trips, but I've come across one or two who are prepared to sleep in their car overnight. Drying wet gear is a major consideration and although I have woken in the morning to find my boots frozen to the van floor and my waders rock hard and impossible to get into, I favour the small motorhome.

There are almost as many ways of sheltering for the night as there are good gold burns and some are more functional than others. The aspiring gold digger is likely to try a few in his or her career.

> Ever panned for 'excrement of the gods' or teocuitlatl? That's what the Aztecs called gold!

Norbie prepares Hungarian chicken paprika in nick Eade's (right) tepee during a meet in 2018.

In for a Penny

Why was there a five cent Newfoundland coin in one of my pans of dirt from the Crom Allt? A local historian discovered there had been a Newfoundlander in Tyndrum in the last century. An Edward I silver penny from another stream was explained by a local when he told me there had been a battle between Scots and English in the valley in the 17th century. A round smooth metal disc I took from a crack in Suisgill granite turned out on later closer inspection to be a Queen Victoria 1890 silver sixpence. Your explanation as to why it was there is as good as mine. Old copper pennies are not uncommon and Mike Jones told me that old panners used to plant a coin to mark the end of their working. This had the benefit of showing if it had been disturbed. Johnny Marr would leave a pan overnight in his for the same reason.

Then there was a diamond from the Kildonan. Admittedly it was attached to a piece of metal that would go through a lady's ear. John Redpath had panned it just above the falls and it was pronounced genuine by Jamie Shepherd and myself as it showed a small inclusion when examined under the hand lens.

One panner's unusual find had an even older origin. Leon Kirk had given a relative beginner on the Mennock a few tips. He had kept in touch. On one occasion, on an impromptu visit, Leon was greeted with the words 'Look what I panned yesterday'. He was handed a very well preserved bronze spearhead. They thought it had been lost on a spear fishing trip thousands of years ago. It was given to the National Museums of Scotland in Edinburgh and after examination by an expert and listing in the 2007/08 Treasure Trove in Scotland Report it was given to Dumfries Museum. Estimated at 3,500 years old; it could have been an offering placed in the river.

For sheer size, the largest artefacts that I have come across in a gold burn were huge iron bars with nuts and bolts on the ends, some of them joined by these links. They lie in and beside the Suisgill below its sharp bend. I could not think they were anything other than items left over from the gold rush, but what? I contacted Dr R. M. Callender with photographs and he was as perplexed as I was. His one suggestion was that it might have been part of a bridge structure, as the Duke of Sutherland had been keen to bridge the Suisgill. I still cannot conceive how they were to be used.

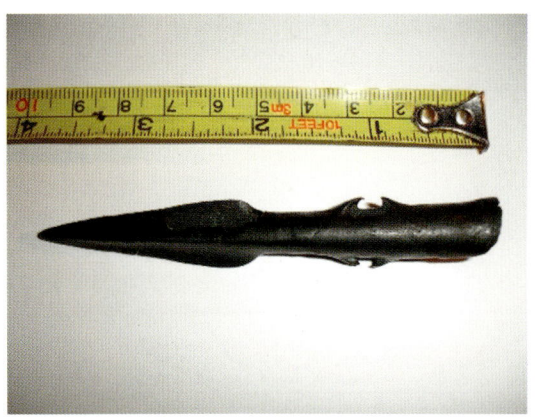

Bronze spearhead, circa 3,500 years old.

Coins not infrequently turn up in the pan. This silver coin was found in a Perthshire stream.

It Remains To Be Seen

Since you cannot make a claim to a piece of gold-bearing ground in Britain, claim jumping is not an issue. Nonetheless, people have been known to move in on a productive spot before the digger is finished. I heard that 'Klondike' Dave (Dave Gray) had turned up to a hole he had spent several days developing to find a younger person in it. Dave asked him to move since he was not finished, but was told to go away. You couldn't meet a more considerate panner than Dave who has given away lots of gold for charity. Dave decided not to make an issue of it. The incident highlights a problem. When do you consider a hole in the gravel abandoned? Few fill in their hole before final departure, knowing the next spate will fill it in quickly. In April 2016 at a meet at Baile an Or a few of us were surprised to see piles of stones at one or two spots. On enquiry at the end of the day, Nick Eade informed us that it was a recognised way of saying you were not finished digging there. Whether it becomes adopted remains to be seen, but if respected it is a good idea.

The stones indicate the panner is still working on this spot.

Dan Haddow's brother Rory was given two pieces of advice by his more experienced sibling. The first was that he should not dig in the bank. The second was that he told no-one of the location of any nuggets he found. Armed with this, Rory went off to see what he could find. At the end of the day he had been successful. He had found a small nugget. "Where did you get that?" asked an astonished Dan. Rory promptly reminded him of rule two.

Let's Dig Deeper

Most panners give little heed to whether the gold they are finding is coming from a mesothermal lode, a porphyry, a breccia pipe, is epithermal or stratabound gold. Most, however, will be aware of the idea that hydrothermal solutions have brought up the gold from deeper parts of the earth. It is also now becoming common knowledge that the earth is rich in gold but that 99% of it is out of reach in the core. The idea that an exploding star is probably the source of this heavy metal and it was this space dust which provided part of the material used to form the earth is in the common domain. More recently, colliding neutron stars are thought to have given the forces needed to make heavy metals. The gold in the crust is now thought to have come from gold-rich meteor bombardment 200 million years after the earth was formed. This did not penetrate below the level from which the silica-rich hydrothermal gold-bearing solutions could be released to nearer the surface. Millions of years of erosion have exposed some of these veins.

From time to time more serious discussions take place in panning circles regarding the source of the alluvial gold, its shape and the speed of replacement, if any. In fact the BGS have a map which classifies Leadhills/Wanlockhead gold as alluvial. At the same time in a paper from the School of Earth and Environment of the University of Leeds, freely available on the internet, headed, 'Research into indigenous gold mineralization', a single grain was stated to be 'coated on one side by layers of fine sugary quartz and a fine grained sedimentary rock'. This rock decomposed over six months when stored in water. "This suggests that the relatively high concentration of gold in this part of the river is due to the liberation of gold grains from the local rock, and the rock itself quickly disintegrates to leave the thin gold grains." The grain was described as 'flakey' but a photo suggested it was also river-worn. The river in question was the Mennock Water. The find was being cited to support the theory that alluvial gold does not travel far from its source. Further work on the silver content in gold from two separate sites on the Mennock showed substantial differences. This further supported the contention that the alluvial gold had come from two separate sources, but had not travelled far. However, later in the article it was mentioned that the silver content can vary within the same mineralized structure. Nonetheless the work on the Mennock had used median scores and one area had twice the silver content on average than the other. Panners do talk about the carat of the gold from different streams and some get samples assayed. The Mennock is frequently quoted as being the purest of Scottish gold and I've heard the figure of 22 carat mentioned, but there is scientific evidence that even within a stream the purity will vary. This work on the micro-chemical signature is beyond the resources of the gold panner. In a conversation with Dr. Rob Chapman at Leeds University, after I had offered to send him some samples, he said that the latest equipment they had installed was proving less useful than their previous set-up. He was snowed under with backlog. There are many other minerals which can be used to give a characteristic signature to gold. This research is being used in archaeology to find where our ancestors were sourcing gold for the ornaments which turn up from time to time. With very expensive equipment involved, what hope is there for the panner to unlock the secrets of gold sources and movement? There are, however, other pointers where close observation can give food for thought.

Any samples of Scottish gold that I have seen in a quartz vein are jagged. I have come across alluvial gold with almost this morphology in two burns and thought, 'that has not moved far'. The article mentioned above takes up this idea. It says that many studies have been done into the relationship between gold grain morphology and distance transported. Its conclusion is that there are a number of variables which make it difficult to come up with a formula. The variables include the original grain size and the 'energy of the fluvial environment'. The auriferous quartz vein my wife found had the largest pieces already smoothed down where they formed part of the stream bed, but inside the morphology was totally different. I still like to think that if I find very rough gold, I'm close to its source, but so far, I've not seen another source vein.

Although the panner may not give much thought to the minerals contained in gold, there is a saying 'Gold rides an iron horse'. German miners referred to 'eisenhut' or iron hat. It alludes to the fact that gold and iron are often associated, the latter frequently in the form of iron sulphate, i.e. iron pyrite, commonly known as fool's gold. When it decomposes a red staining can be found in the rocks and gossan, the weathered rock material on and near the surface of an ore body. It was this that alerted an Irish geologist to the mineralization and possible gold potential of the Dalradian rocks in Northern Ireland. There are very few streams where you get gold without considerable evidence of iron in some form or other.

Other aspects of this more scientific nature include the extent to which streams are being recharged with gold. Vince Thurkettle classifies streams into two types: those which are getting replenished with gold and those which are not. Of the former, perhaps the only good examples are in Merionethshire where the Mawdach for one is surrounded by gold veins and former gold mines. Most streams are in the latter category. We have also had a conversation as to how gold can bury itself in cracks to the point it cannot be seen. Above the water I could see how freeze and thaw would open and close cracks, allowing the gold to drop out of sight. Would this work as well below the surface? Vince has another theory. Could the vibration caused by moving boulders open the cracks momentarily? A keeper I spoke to in Perthshire described the noise created by moving boulders in a spate in a burn near his home. I have also heard the noise of gravel movement in the Spey in flood. The vibrations caused could help gold to sink well into cracks.

Is any of this going to help the amateur gold seeker? Perhaps, but the best way to treat gold fever and to enjoy the various pleasures of looking for gold is to get out there and dig.

Does our hobby have a future in Britain?

'Gif ony myne of golde or siluer be fundyn in ony lordis landis of the realme and it may be prowyt that thre halfpennys of siluer may be fynit owt of the punde of leide. The lordis of parliament consentis that sik myne be the kingis as is vsuale in vthir realmys.'

In February 2018 I received a call telling me to read an email forwarded to me. It had been sent to Leon Kirk in his role as owner of Goldpanning Supplies UK. In essence it told him to cease selling bags of paydirt which included Crom Allt gold. It came from the Crown Mineral Agent on behalf of Crown Estate Scotland and ended with the thinly-veiled threat 'that should your activity continue then Crown Estate Scotland will consider what further options are available'. The email also stated 'the Crown Estate Scotland has not given you or your company permission to take away or sell any of the Mines Royal contained in the pay dirt that you sell'. Within a week, offers to sell Scottish gold were disappearing from the internet. Leon had removed his paydirt which was on sale only because of a requested demand. It was more trouble than it was worth.

What are Mines Royal and what gave the Crown Mineral Agent the authority to make these statements? The answer lies in part in the quote below the title. It is the oldest law on the statute books in Scotland and dates back to 1424. A more coherent version reads, 'Item, of any gold and silver mines that are found in any lord's lands of the realm, and it may be proved that three halfpennies may be refined of silver from a pound of lead, the lords of the parliament consent that such mines shall be the king's, as is the custom in other realms.' How did a law aimed at giving a monarch the right to any mines with gold or silver come to be used against hobbyists who fossick for a few flakes of gold and if they are lucky find the odd nugget. What appears from all the literature is that the term Mines Royal has become synonymous with gold and silver. In a 1995 letter from the Crown Mineral Agent in response to a request to prospect in the Ladder Hills I was informed of the following; 'You are probably aware that all Mines Royal, that is gold and silver, are vested in the Crown Estate'. It went on to say 'A letter of no objection is normally issued to an individual undertaking exploration as a hobby...' After a correct presumption that my prospecting was small scale and pointing out that I had to seek access permission they said they had no objection.

Today the statement on the Crown Estate website is not so accommodating. Under the heading GOLD PANNING it states 'As a responsible mineral owner, we discourage gold panning as such activity causes harm to the aquatic environment and damages wildlife and their habitats in and around streams and rivers. Because of the damage gold panning causes, we do not grant permission for people to remove any gold. The permission of the landowner is also required to gain access to the stream or river to pan for gold. It is an offence to remove gold found by recreational panning without the relevant permission of the owner.'

It then refers the reader to various environmental websites to check on the alleged damage, but not to a Scottish one. Scottish Natural Heritage had a policy but this was withdrawn, though the Scottish Environment Protection Agency had an enlightened one towards panning.

The Crown Estate does not have a right to gold and silver mines in all of Scotland. Historically, the right to gold and silver in areas in Sutherland and around Leadhills was granted to the landowners. There are other smaller areas.

Gold panning as a hobby now receives a great deal of attention. Adverse publicity has made landowners who do have the right to allow panning think twice about granting permission. Clearly-stated rules were drawn up for Kildonan and the Mennock. It doesn't help that there are those who through ignorance or more likely arrogance (and they will know who they are, as do most other panners) have openly flouted the rules, thus putting the hobby at risk. Selling Kildonan gold, digging into the banks, using a banned sluice and digging in spawning grounds in the wrong season and leaving litter are five instances, but there are more.

One solution to continuing this absorbing and therapeutic hobby would be to take up looking for platinum, ilmenite and magnetite instead. They do not get mentioned in Mines Royal. If you inadvertently found gold you could, like anglers, practise catch and release.

Alternatively, seek permission, respect the rules and the environment and continue to enjoy looking for gold.

GLOSSARY

batea An open cone-shaped gold pan.
bedrock The solid rock below loose deposits. It is not always visible in streams.
boot scooping Moving gravel back and forth with the foot to allow gold to sink from the upper layers and concentrate it over the bedrock. If done correctly, at the right spot, gold free gravel can be quickly removed leaving a rich concentrate.
bulb A suction hand tool for lifting gold from under water. Beloved of snipers, it is comprised of a large rubber bulb and metal tube.
carat A measure of the purity of gold (see Appendix 2).

clean up	The process of emptying a sluice to remove the gold.
colour	A term used to donate bits of natural gold.
crevicing	Taking gold from bedrock cracks by scraping then sucking up the loosened material.
cupel	A small vessel used for refining and assaying gold. It is made from crushed calcium carbonate which can be mineral or organic in origin.
cupellation	The process of separating gold and silver from base metal impurities such as antimony, arsenic and bismuth to name three. Under intense heat the impurities are partly vaporised and oxidised and absorbed into the pores of the cupel.
drysuit	A totally waterproof body covering leaving only hands and head exposed.
flake	A small piece of gold in which the third dimension is very limited.
fluvioglacial deposits	Sands and gravels deposited on, beside, under or in front of a glacier by its melting waters.
gravel monkey	A derogatory term for a civil engineer and or geologist but here used to refer to the individual asked to do the gravel digging for a course of beginners.
hardpack	Gravel which has been *in situ* for a long period and has become 'cemented' by a mineral such as iron.
heavies	Minerals of a higher specific gravity which are left in the pan or sluice near the end of the sorting process.
light gold	A term used to refer to very thin flakes of gold which never weigh as much as you think they should given the area they cover in the pan.
miner's moss	A plastic open carpet used to trap gold in a sluice.
moss washing	Stripping moss from stream rocks to recover fine gold filtered out of flood water.
noble metals	Those metals which do not oxidise or tarnish in air, including platinum, gold and silver.
nugget	A lump of gold. In Britain anything around or over a gram. A much misused word as in panning competitions it is used to describe tiny flakes. Also much abused on eBay.
pay dirt	Any naturally ocurring loose deposit which allows the prospector to accrue gold.
picker	A piece of gold capable of being lifted easily between the finger and thumb, but smaller than a nugget. It has a greater third dimension than a flake.
placer	A deposit of sand or gravel containing gold.
riffles	The bars in a pan or sluice which help trap gold.
shuftyscope	A handheld glass bottomed viewing pipe enabling subsurface inspection.
sluice	An open ended elongated box using riffles and or mat to separate gold from gravel by the use of flowing water.
sniping	A gold recovery method where individual bits of gold are spotted, usually in bedrock cracks, then removed, normally with a bulb. It is usually done with a drysuit, snorkel and mask, but can also be done in waders with a shuftyscope or viewer.
snuffer bottle	A plastic bottle with a nozzle which is used to suck and temporarily store gold from a pan.
speck	A small piece of gold not worthy of being called a flake.
trommel	A perforated rotating drum used to separate large rocks from finer gold bearing material on a commercial mine.
vein	A seam of a different mineral running through bedrock.
vial or phial	A small glass or plastic bottle used by panners for keeping and displaying gold.

APPENDIX 1

Most contemporary British panners, measure their gold in grams. Some have amassed enough to have vials with ounces in them. There are other units which were commonly used in the past and occasionally turn up in historical accounts.

The ounce used in gold weighing is the troy ounce. At the time of the Kildonan Goldrush, returns were recorded in grains, pennyweights and troy ounces. In some ways it is an easier system to remember.

1 troy ounce = 20 pennyweight = 480 grains
1 pennyweight = 24 grains

Use of the gram but retention of the troy ounce has made for an awkward figure to remember if you wish to be accurate.

1 troy ounce = 31.103 grams (most people round it to 31 grams)

Most modern scales will change from grams to grains at the flick of a switch. Using the grain for small amounts gives you the satisfaction of larger whole numbers since 1 grain equals 0.065 grams. I'm not advocating a change to the old system, but measuring in grains does boost the ego.

A word of warning. The abbreviation for a gram is 'g', for a grain it is 'gr'. Soon after the news of the big nugget was broken to the press on the 26th July 2018, a colleague of Dr Neil Clark of the Hunterian Museum of Glasgow University was emailed by a curator at the Natural History Museum. The curator was wondering what all the fuss was about as they had a nugget from the Suisgill that weighed 128 grams. It turned out the nugget had been on display for years and was part of the Russell mineral collection. Neil questioned the weight saying 'That surely a nugget of that size would have been mentioned in the literature somewhere'. Neil had been responsible for bringing together a collection of some of the biggest British nuggets for the summer 2014 exhibition on gold in Glasgow University's Hunterian Museum and it bothered him that this one had escaped his attention. The curator went to have a look and noticed straight away that it was not big enough to be 128 grams. It turns out that the original label had said 128gr, which was misinterpreted as grams rather than grains. The nugget was around 8 grams. Apparently, Neil informed me, this was not the first misreading of grains as grams in museum history.

At a British Goldpanning Association meeting soon after, Neil left a few members gobsmacked, including Leon Kirk who had fronted the release of the news of the biggest nugget for 500 years. He told them initially that the Russell collection's Suisgill nugget was said to be 128 g. I wasn't there, but I can picture the smile on Neil's face as he relished the telling of the full story.

APPENDIX 2: PURITY

The purity of gold is a topic more often discussed in jewellery shops than in panning circles, but more recently, with more information on the topic, figures from different streams are bandied on the riverbank. The purity of gold is measured in carats. Pure gold is 24 carats. The word carat is derived from various words referring to the fruit of the carob tree. These seeds were used as weights in Middle East markets. Confusingly, the word carat is also used as a measure of weight for gems being 0.2 of a gram. There is another way of quantifying the purity, called fineness, expressed in parts per 1,000. This system is more likely to be used in gold jewellery. The table below shows the relationship between the two measures. To get a percentage of gold, move the decimal point in the fineness figures one place to the left. Hallmarks on gold jewellery round the fineness numbers to those shown in the table.

Carats	Fineness	Hallmark
24	999.9	999
22	916.6	916
18	750	750
14	583.3	585
9	375	375

Below 18 carats is considered an alloy.

Since in nature gold has inclusions, your flake or nugget is not going to be 24 carat. The most likely mineral reducing the purity is silver, but others include copper and mercury. Some Mennock gold has as little as 5.6% silver with very little other minerals. This would make it 22 carat. Thus Mennock gold is often quoted as the purest in Scotland. Other gold in the Leadhills, Wanlockhead area has up to 12% silver. This would be classed as 18 carat. The silver content of gold grains from the same mineralized structure can vary, but significant variations in fineness in a sample of river gold might suggest it came from more than one source.

It is possible to buy kits which by using different strengths and combinations of acid will give bands of purity. Again they are for the jewellery trade more than the prospector. There are also hand-held machines which will do the job, but they come at a price beyond the pocket of most amateur prospectors.

APPENDIX 3: THE EFFECTS of MOVEMENT on GOLD

Gold is a soft malleable mineral and its resistance to physical weathering is low, i.e. the sharp-edged often spiky mineral that is seen in quartz samples becomes pummelled into a smoother round edged flake or nugget when released into a dynamic environment. I can recall one instance in Perthshire when I thought the contents of my pan had not moved far from where they had been released from their matrix. On the other hand some Kildonan nuggets were like pills. Geologists have tried to classify the morphology of gold grains for use in source exploration and identification. Admittedly, at times, the size of the bits being classed would be missed by the average amateur, but the idea applies to visible pieces.

R.N.W. DiLabio's classification for grains from till is as follows:

Class	Shape	Surface texture
Pristine	Block Rod or wire Leaf Crystal Star Globule or bleb	Smooth surfaces Angular edges Grain moulds clearly visible Thin edges not curled Some striae
Modified	All pristine shapes damaged, but visible.	Leaf edges and wires bent, curled blunted and clubbed edges, grain moulds preserved where protected, moderately striated, felty texture where damaged.
Re-shaped	Well rounded grain outline, folded rod, wire, flake, rounded block, typical discoid placer flake, nugget.	Porous, scaly, felty or spongy, rarely striated.

Many of these terms may be unfamiliar but some give the gist of what happens during transportation. Despite many studies into the relationship between morphology and the distance from source, other factors such as the original size of the gold and the energy of the transport system make it difficult to draw up a definitive table.

On the other hand, gold is very resistant to chemical weathering. Therefore, although its shape may be altered with transportation, any suite of other minerals in it will not change. At this point we are entering the realms of the geochemist who through the use of various forms of electronic scanning can analyse the chemical signature of a gold grain. All the non-gold minerals in a grain and their percentages can be tabulated. This microchemical characterisation gives a gold particle signature. This is basically like a fingerprint and is related to the fineness. Most panners will not give it much thought when digging. Nonetheless there is one rule from this academic work which might interest the amateur prospector hoping to find gold in bedrock. In upland Britain and Ireland the degree of transportation from bedrock source to alluvial occurrence is usually very small.

APPENDIX 4: GOLD IN SEDIMENT SAMPLES

The earliest way of prospecting for gold has not changed much in thousands of years and is roughly the method employed by panners today. To find new bedrock sources for commercial mining, however, surveys based on the chemical analysis of stream sediment samples have supplanted it. This is based on the logical premise that stream sediments are composed of the products of erosion and weathering from the drainage basin upstream. They will therefore reflect the bedrock geology and the drift of the catchment area and thus will contain any minerals in that area. The survey still involves manual work to get the samples. Only when analysed in the laboratory will it be seen whether there is potential for further exploration.

There are guiding rules as to where and how to sample in order to negate the effects of pH (water acidity) and Eh (electron activity) on samples as well as the effects of human contamination. The British Geological Survey (BGS) has undertaken two major surveys in the last three decades. They are the Mineral Reconnaissance Programme (MRP) and the Geochemical Baseline Study of the Environment (G-BASE). The reports and maps of these can give useful information for the amateur prospector. In some, the amount of gold (Au) is mentioned in parts per billion (ppb). In others like the geochemical atlases of various regions, maps of individual elements give values as a single black line, or on a coloured isoline map. Gold has not been recorded, though pathfinder minerals have. The large maps contain information from samples taken at an optimum density of 1 per 1.5 to 2km^2. Since the sieves have a mesh of 2mm and 150 microns ($^7/_{20}$ths of a mm), it is only the finest material that is taken. The less than 150 microns sample is left to settle in a fibreglass pan and is decanted into a sample collection bag. The excess sediment which has gone through the 2mm sieve is then panned to collect the heavy mineral concentrate. This work has been undertaken by Earth and Geo-science students during their summer vacation.

Scotgold Resources' 2010 report has a map which shows gold amounts from sediment samples in an area of the SW Highlands. It also uses the unit ppb and categorizes parts of streams in a five division scale. The lowest 5 rows of the table below show the categories and an assessment of the likelihood of finding gold in your pan.

Au in parts per billion gold in a sediment sample – ppb.

0.012	Approximate gold content of seawater – too dilute to extract commercially
0.02	Content in fresh water
5.0	Average concentration in the earth's crust
5 – 50	Background content of all streams – little chance of visible gold
51 – 100	Not worth looking at – might find a trace if figure inaccurate due to sampling limitations
101 – 1,000	Should be able to find visible specks with effort
1,001 – 10,000	Well worth looking at – some pockets could be quite rich
10,001 – 22,000	These streams should provide grams and ounces over time

The figures used by Scotgold are based on the British Geological Survey's sediment sampling in its licence areas.

BIBLIOGRAPHY

Heddle, M. Forster, *The Mineralogy of Scotland, Volume 1*, David Douglas, Edinburgh, 1901.

Landless, J.G, *A Gazetteer to the Metal Mines of Scotland, Occasional Paper No.1*, The Wanlockhead Museum Trust, 1985.

Adamson, G.F.S., *At the End of the Rainbow, The Occurrence of Gold in Scotland*, Goldspear (UK) Ltd., Beaconsfield, 1988.

Clark, Neil D.L., *Scottish Gold – Fruit of the Nation*, Neil Wilson Publishing Ltd., in conjunction with The Hunterian, University of Glasgow, 2014.

Callender, R.M., *Gold in Britain*, Goldspear (UK) Ltd., Beaconsfield, 1990.

Porteous, J. Moir, *God's Treasure-House in Scotland – an abstract*, Goldpanners Association, 1876.

Callender, R.M. & Reeson, P.F., *The Scottish Gold Rush of 1869*, Northern Mine Research Society, Sheffield, 2008.

Bancroft, Caroline, *Colorado's Lost Gold Mines and Buried Treasure*, Johnson Books, Boulder, 1961.

Bishop, Tom, *Gold! The Way to Roadside Riches*, Johnson Books, Boulder, 1971.

Wendt, Ron, *Where to Prospect for Gold in Alaska Without Getting Shot*, Goldstream Publications, Wasilla, 1956.

Service, Robert, *Songs of a Sourdough*, Ernest Benn Ltd., London, 1956.

Gilliland, Ellen, *Summit – A Gold Rush History of Summit County Colorado*, Alpenrose Press, Silverthorne, 1980.

Meyer, Kathleen, *How to Shit in the Woods: An Environmentally Sound Approach to a Lost Art*, Ten Speed Press, Berkeley, 1989.

Macfie, H. and Westerlund, H.G., *Wasa Wasa*, Allen and Unwin, London, 1953.

Minerals in Britain: Past Production… Future potential – Gold, British Geological Survey / DTI, Keyworth, Nottingham. 1999.

Geological Survey Ten Mile Map, North Sheet, Third edition, Institute of Geological Sciences (BGS), 1979.

Bedrock Geology, UK North, 5th edition, British Geological Survey, Keyworth, Nottingham, 2007.

May the fever ever be with you.